DATE DUE

MY 28 '98			
JY 30 '98			
NO 10 '98			
DE 1 '98			
FE 17 '99			
AG 5 '99			
MY 21 03			

DEMCO 38-296

Spaceflight

A Smithsonian Guide

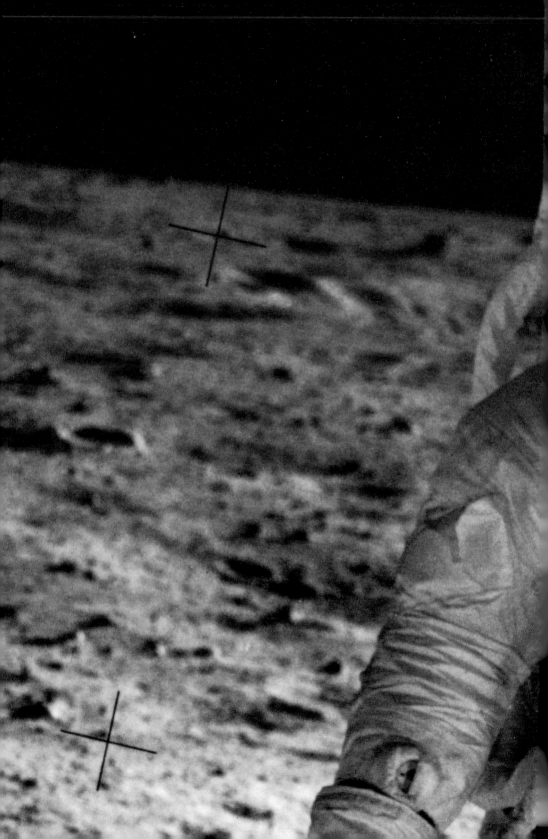

Spaceflight

A Smithsonian Guide

Valerie Neal
Curator for Skylab, Space Shuttle,
and Spacelab
Department of Space History
National Air and Space Museum

Cathleen S. Lewis
Associate Curator
Aeronautics Department and
Department of Space History
National Air and Space Museum

Frank H. Winter
Curator of Rocketry
National Air and Space Museum

Macmillan • USA

Macmillan • USA
A Prentice Hall Macmillan Company
15 Columbus Circle
New York, NY 10023

A Ligature Book

Copyright © 1995 by Ligature, Inc.

...rrent resources.
...o: Ligature, Inc.,

...ansmitted in any form
...g, recording, or by any
...riting from the Publisher.

pages 2–3. The first step to the Universe: the Moon.
pages 4–5. The Space Shuttle rises into the sky.
pages 6–7. The first men on the Moon: Neil Armstrong and the Lunar Lander reflected in Edwin Aldrin's visor.
page 8.   A Gemini spacecraft floats high above the Earth.
page 11.   The Magellan space probe launched from the Space Shuttle payload bay.

Library of Congress Cataloging-in-Publication Data

Neal, Valerie.
  Spaceflight: A Smithsonian Guide / Valerie Neal, Cathleen S. Lewis, Frank H. Winter.
    p.   cm.
    "A Ligature book."
    Includes bibliographical references and index.
    ISBN 0-02-860007-X.   ISBN 0-02-860040-1 (pbk.).
    1. Space vehicles—History. 2. Outer space—Exploration—History.
    3. Space flight—History.
I. Lewis, Cathleen, S., 1958–   II. Winter, Frank H. III. Title.
TL795.N43   1995
629.4'09—dc20   94-23448 CIP

Ligature Inc.

Publisher	**Editorial**	**Production**	**Contributors**
Jonathan P. Latimer	Susan Judge	Anne E. Spencer	Carol Sutton
Series Design	Mary Ashford	Ron Frank	Dick Bolster
Patricia A. Eynon	Robert Costello	Paul Farwell	Alice Holstein
Design	Susanna Brougham	Marguerite Meister	Holly Manry
Julia Sedykh	Barbara Simons	**Research**	
	Elizabeth A. Mitchell	Michael Pistrich	
	Nancy Ludlow	Kristen Holmstrand	
	Steven Thomas	Carolyn Mitchell	

The Smithsonian Institution

Macmillan

Patricia Graboske	Robert Curran	Michael J. Neufeld	Robin Besofsky
Jim Wilson	Matthew A. Greenhouse	Howard A. Smith	
Bruce A. Campbell	Kristine Kaske	Steven Soter	
Robert A. Craddock	Melissa Keiser	Priscilla Strain	
Timothy J. Cronen	Lillian D.Kozloski	Karen Whitehair	
Tom D. Crouch	Allan A. Needell	James R. Zimbelman	

Macmillan books are available at special discounts for bulk purchases, for sales promotions,
fund-raising, or educational use. For details, contact:

Special Sales Director
Macmillan Publishing Company
15 Columbus Circle
New York, NY 10023

Manufactured in Hong Kong
10 9 8 7 6 5 4 3 2

Contents

The Elements of Spaceflight

When Galileo trained his telescope on the skies in 1609, a thin veil was lifted. What had appeared to the human eye for centuries as a decorative dome became a vast universe of real worlds to explore. Gradually, engineers and scientists became convinced that the barriers to spaceflight could be overcome.

The Space Shuttle
Columbia, mounted
on its mobile launcher
at the Kennedy Space
Center in Florida, is
readied for its first mission in April, 1981.

In 1967 Apollo 4 was propelled more than 11,000 miles above Earth, burning 15 tons of propellant each second. This mission tested the conditions of reentry for the new Apollo Command Module that would carry astronauts to and from the Moon the following year. An imaginary rocket (inset), illustrating a Jules Verne novel, foretold twentieth-century spaceflight.

Dreams of Spaceflight

Myths and tales of flying to the Moon or the stars are probably as old as humankind. One of the earliest accounts recorded was by a Greek named Lucian of Samosata. In A.D. 160, Lucian wrote a story of a voyage to the Moon, *True History (Vera Historia)*, in which the hero's sailing ship is caught by a whirlwind and taken on an eight-day journey to the Moon. Lucian also wrote a satire entitled *Icaromenippus* (a play on words that combines the name of the hero of the satire, Menippus, with that of the famous mythological flier, Icarus). Menippus reaches the Moon and then continues to Olympus, where Zeus is angered by the intrusion and orders Mercury to take away Menippus's wings and return him to Earth: the story is a less than auspicious beginning for human spaceflight.

Birds were often the preferred means of flying in early imaginary trips to the Moon and beyond. In the ninth century, a Persian poet described flying to the Moon on a throne pulled by eagles. In 1638, Bishop Francis Godwin wrote *Man in the Moone*. Godwin's voyager traveled to the Moon in a machine pulled by wild geese.

Johannes Kepler, the German astronomer who first worked out the mathematical shape of the orbits of the planets, imagined spaceflight in his book *Somnium (Dream)*, published in 1634 after his death. Kepler speculated about the effects of weightlessness, including the incorrect suggestion that a region of weightlessness exists between the Moon and Earth where the gravities of the two bodies cancel.

In the mid-1600s, Cyrano de Bergerac, the French poet and swordsman, wrote two novels about travel to other worlds. In one, Cyrano describes traveling in a "flying-chariot"—a box propelled by rockets attached to it.

Yuri Gagarin became the first person in space April 12, 1961. Wearing a pressurized suit, the Soviet cosmonaut rocketed into orbit aboard Vostok 1.

Two hundred years later, another French author, Jules Verne, wrote science fiction that included *From the Earth to the Moon* (1865) and *Around the Moon* (1870)—works in which Verne's characters travel away from Earth inside a huge projectile fired from a giant cannon. Many of Verne's ideas were scientifically correct and seem almost prophetic today. For example, his book's spacecraft was launched in Florida so that the speed of Earth's spin near the equator would help it escape Earth's gravity. That is one of the reasons that Cape Canaveral was selected as a launch site nearly 80 years later.

H. G. Wells also helped popularize the idea of spaceflight. Wells's *The War of the Worlds,* published in 1898, graphically depicts an invasion of Earth by Martians who look like pale octopuses. Later, in *The Shape of Things to Come,* Wells tried to peer a hundred years ahead in time. His predictions included the launching of a piloted space capsule by a space cannon similar to the one proposed by Verne.

After World War I, interest in space travel led to accounts in newspapers, comics, magazines, and on the movie screen with characters such as Flash Gordon and Buck Rogers. World War II saw the introduction of large rockets with real warheads, and postwar science fiction became more sophisticated. The modern science-fiction story used technology as a field of speculation, sometimes with spectacular results.

In 1945 the modern science-fiction writer Arthur C. Clarke proposed using satellites to relay communications to different parts of Earth—more than 20 years before the first satellite was launched! His other writings described travel to the Moon in *Prelude to Space* (1951), and to Mars in *The Sands of Mars* (1951). The movie *2001: A Space Odyssey,* made in 1968 from another of Clarke's novels, is still one of the best available depictions of real spaceflight.

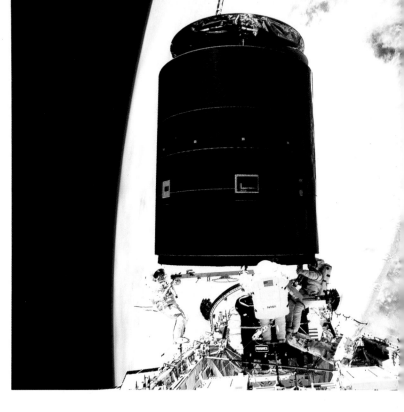

Astronauts Richard J. Hieb, Thomas D. Akers, and Pierre J. Thuot perform an extravehicular activity (EVA) in May, 1992. Intelsat VI—a communications satellite of the International Telecommunications Satellite Organization—is captured by the crew and moved toward the cargo bay of the Space Shuttle *Endeavour.*

The Environment of Space

Although people have dreamt about going to the Moon and beyond for many centuries, it was not until the first decades of the twentieth century that the idea of spaceflight became a real possibility. The development of airplanes and of more powerful means of propulsion (engines and rockets) made space travel seem attainable. Scientists were also gathering information about space itself—an environment completely unlike our familiar world.

Space is not only different from Earth, it is lethal to earthly life unless special precautions are taken. To survive you must be protected from the vacuum of space, compensate for weightlessness, deal with wide variations in temperature, and shield yourself from intense radiation. You also have to bring along all the oxygen, food, and water you will need to survive while in space.

The Vacuum of Space In space there is simply no air. There are a few atoms and molecules of hydrogen and other elements, but so few that space is virtually a vacuum. To breathe, humans must carry their own air supply and stay inside a pressurized, airtight container—a spacecraft or space suit. The vacuum of space has other effects as well. Materials that we rely on every day on Earth behave in unexpected ways. Liquids evaporate very rapidly.

Lubricants such as oil or grease, which work smoothly on Earth, thicken and harden in space. And, because there is no air, there is also no atmospheric friction, or *drag*.

Weightlessness This state is also called *freefall, zero-gravity* (or zero-g), or *microgravity*, but it is not the absence of gravity. In fact, gravity is what keeps the spacecraft and everything inside it in orbit. Being in orbit means that the force of gravity down (toward Earth) is exactly balanced by the centrifugal force of rotation up (away from Earth). Thus, a traveler in space feels no "weight," but still has normal mass. In a state of weightlessness, walking is impossible; instead, you "float" from place to place. Objects that would be too heavy or too large to move on Earth can be easily handled. For example, a single astronaut can maneuver a satellite that weighs hundreds of pounds on Earth.

Weightlessness can cause temporary nausea and, on extended flights, loss of muscle tone and bone strength. It also affects the heart and circulation, which are not working against gravity. Blood and other bodily fluids tend to collect in the upper part of the body, as indicated by the puffy appearance of the faces of some space travelers. Special exercises and equipment have been developed to keep their heart and other muscles in good condition during space missions.

Making her first space-flight in September, 1992, astronaut Mae C. Jemison works in the science module aboard the Space Shuttle *Endeavour*. Jemison, with five other NASA astronauts and a Japanese payload specialist, conducted research under microgravity conditions on the Spacelab-J mission—a joint effort between Japan and the United States.

Space Shuttle *Columbia*, carrying a crew of seven and the Spacelab Life Sciences–1 laboratory, lifts off on June 5, 1991. Focusing solely on life sciences research, the crew explored the physiological changes that occur during spaceflight, and the consequences of the body's adaptation to microgravity and readjustment to gravity upon return to Earth.

Temperature The effect of sunlight heating an object in space is much more intense than when the object is protected by Earth's atmosphere. On the other hand, the temperature of space itself, which is near absolute zero (-459° F or -273° C), rapidly cools an object. This means that the sunny side of an object can be hotter than the temperature of boiling water, while its shaded side may be more than 100 degrees below freezing (-73° C). It also means that when an object passes in or out of Earth's shadow, its temperature can change significantly in a matter of minutes. Spacecraft are usually carefully insulated and have protective coatings to help regulate temperature.

Radiation The Sun and other stars produce cosmic rays and other high-energy particles and radiation. These can damage delicate electrical equipment on a spacecraft and disrupt power flows. This radiation is also hazardous to humans, even when inside a spacecraft.

Unfiltered sunlight also contains harsh ultraviolet light. Humans need sun filters to protect their eyes and help them see better. Unfiltered ultraviolet light also causes some materials, such as rubber and plastic, to decompose.

Magnetic Fields Earth and some other planets have strong magnetic fields, as does the Sun. These fields can disrupt delicate instruments and affect the navigation of a spacecraft.

Transportation of the Soyuz launch vehicle and Soyuz spacecraft (left) afford a view of five rocket clusters. The individual engines together provide the thrust necessary to propel the Soyuz into space.

Designing a Spacecraft

Although the design of a particular spacecraft is dictated by its specific mission, the general design of all spacecraft must overcome the same problems. First, the spacecraft must have a propulsion system that provides enough prolonged thrust for it to escape Earth's gravity and deliver its *payload*, meaning passengers or instruments, to its destination. Second, the design must take into account all the aspects of the environment of space. The craft must be able to withstand the stresses of liftoff and carry enough fuel and supplies to complete its mission and sustain its crew. Finally, some spacecraft must be able to reenter Earth's atmosphere and land safely.

Propulsion

The only engine that produces enough energy to escape Earth is a rocket. Unlike jets or internal combustion engines, which use the oxygen in the air to burn their fuel, rockets carry not only their own fuel but also a substance that supplies oxygen, called an *oxidizer*. Together, the fuel and the oxidizer are referred to as the rocket's *propellant*.

The simplest rockets burn propellant inside a container with an opening at one end to let the exhaust gas escape. As the fuel burns, the hot gases expand rapidly in all directions. If the chamber were closed, this rapid expansion could make it explode. But in a rocket, the exhaust gases escape through the opening in one

The propellant weight usually amounts to 90 to 95 percent of the weight of the whole rocket.

end, called the *nozzle*. This rapid escape of gases creates a force, called *thrust,* that pushes the rocket forward. The nozzle is flared to help control the expansion of the exhaust gases.

The amount of thrust a rocket engine has is usually measured in pounds or tons. Most rockets use up their propellant in the first few moments of a flight. Some spacecraft, such as the Space Shuttle, use small rockets to maneuver in space, but for the most part, a spacecraft coasts toward its destination.

Solid Propellant A solid propellant is a compressed form of fuel and oxidizer that is solid, not liquid or gas. The earliest rockets, built by the Chinese, used a solid propellant: gunpowder. Today, solid-propellant mixtures are much more powerful and complicated than gunpowder. The boosters for the Space Shuttle, for example, use a mixture of ammonium perchlorate, powdered aluminum, and additives.

The shape of the solid block of propellant can affect its performance because the thrust of a rocket depends on the amount of fuel burning at one time. Solid propellants are usually prepared in liquid form, then poured into a mold or the casing of the rocket itself, and allowed to harden. The boosters for the Space Shuttle are made in this way.

The advantages of solid propellants lie in their simplicity and reliability. They are easier to store than liquid propellants, can withstand changes of temperature better, can be readied for

Solid-Propellant Rocket

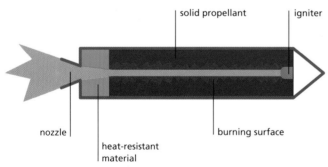

solid propellant   igniter

nozzle   burning surface

heat-resistant material

Liquid-Propellant Rocket

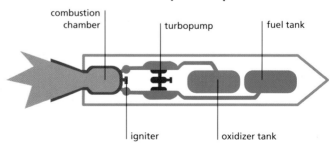

combustion chamber   turbopump   fuel tank

igniter   oxidizer tank

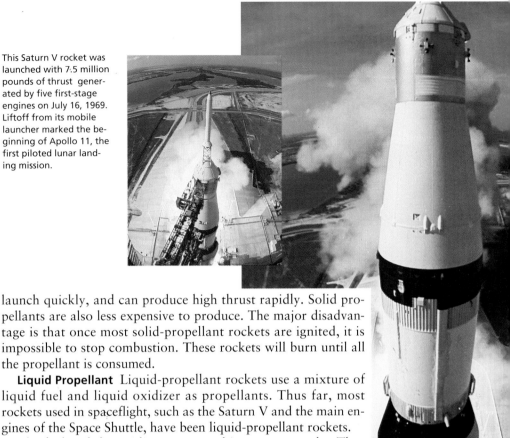

This Saturn V rocket was launched with 7.5 million pounds of thrust generated by five first-stage engines on July 16, 1969. Liftoff from its mobile launcher marked the beginning of Apollo 11, the first piloted lunar landing mission.

launch quickly, and can produce high thrust rapidly. Solid propellants are also less expensive to produce. The major disadvantage is that once most solid-propellant rockets are ignited, it is impossible to stop combustion. These rockets will burn until all the propellant is consumed.

Liquid Propellant Liquid-propellant rockets use a mixture of liquid fuel and liquid oxidizer as propellants. Thus far, most rockets used in spaceflight, such as the Saturn V and the main engines of the Space Shuttle, have been liquid-propellant rockets.

The fuel and the oxidizer are stored in separate tanks. They are pumped into the combustion chamber where they mix, ignite, and burn to produce thrust. Some mixtures require igniters to start the combustion process. Others, called *hypergolic,* ignite immediately on contact with each other and need no igniter. Because the flow of fuel can be controlled, so can the thrust of the rocket. Liquid-propellant engines can also be shut off and restarted, unlike most solid-propellant engines.

Liquid-propellant rockets have been the most common kind used in spaceflight because of their high energy in relation to their weight and the ability to control their thrust. Common liquid fuels include alcohol, kerosene, liquid hydrogen, and hydrazine. Common oxidizers include nitrogen tetroxide and liquid oxygen. Many of these propellants are difficult or dangerous to handle, transport, or store.

Hybrid Rockets Hybrid engines bring together the technology of solid and liquid propellants. In most cases, the fuel is solid and the oxidizer is liquid. These engines combine the simplicity and reliability of solid-propellant engines with the ability of liquid-propellant engines to control thrust and to stop and restart.

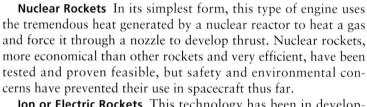

This view of the Earth-orbiting Space Shuttle *Challenger* (right) was photographed with a camera onboard the Shuttle Pallet Satellite (SPAS-01). The liftoff of Space Shuttle *Atlantis* (above) on October 18, 1989, marked the beginning of a five-day mission. *Atlantis* carried a crew of five and the Jupiter-bound space probe *Galileo*.

Nuclear Rockets In its simplest form, this type of engine uses the tremendous heat generated by a nuclear reactor to heat a gas and force it through a nozzle to develop thrust. Nuclear rockets, more economical than other rockets and very efficient, have been tested and proven feasible, but safety and environmental concerns have prevented their use in spacecraft thus far.

Ion or Electric Rockets This technology has been in development since the 1950s. Although it has given good results in the laboratory, ion propulsion has not yet been brought into actual use. In theory, the atoms of a fuel such as mercury or even water are given a positive charge by removing some of their negative electrons. Then a strong magnetic field is used to accelerate these charged atoms and produce thrust. No oxidizer is needed, so an ion rocket can carry more fuel. The thrusts obtained would be relatively weak, but ion rockets could function for a very long time. Because these are low-thrust rockets, they can't be used to launch from Earth. They could, however, be carried by conventional rockets and launched in space, where they would work well. Ion rockets might be well suited to very long flights, such as probes to other stars. Ion systems are still experimental.

The Structure of the Airframe

A spacecraft really has to be able to withstand the rigors of three different environments. First, it must be able to function in the gravity and atmosphere of Earth. Even if the spacecraft will never return to Earth, it still must be built and tested here. Second, the spacecraft must be able to withstand the high heat and strong vibrations of launch and, in some cases, the return to

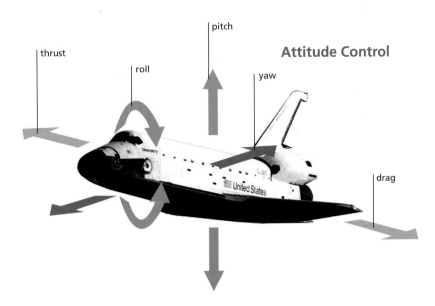

pitch

thrust

roll

yaw

drag

Earth. Finally, it must be able to function in the vacuum of space where it is subject to abrupt changes in temperature and strong radiation and magnetic fields. In some cases, a spacecraft will spend months or even years in operation, so it has to be durable and reliable. It also has to be as lightweight as possible. Each pound of payload weight means many pounds of fuel burned to escape Earth.

Maneuvering in Space

In the weightless, frictionless environment of space, objects still follow Newton's laws of motion. They tend to keep moving in the same direction unless acted upon by some force. Consequently, changing course means restarting an engine or firing another rocket. To be sure that the rocket is fired in the right direction or that its instruments or antennas are pointed at a particular target, the spacecraft must be in the right attitude.

Attitude Control In spaceflight, the term *attitude* refers to the spacecraft's orientation in relation to its direction of motion. On the ground, an automobile, for example, can only move forward and backward along the axis of travel. Turning is a change in its forward motion. However, a spacecraft can move not only forward and backward along the axis of travel, but also (1) from side to side on the horizontal plane, called *yaw*; (2) with its nose pointed up or down on the vertical plane, called *pitch*; and (3) rotating around the axis of flight, called *roll*. Most spacecraft use very small liquid-propellant rockets or pressurized gas-jet thrusters to control their attitude in space.

Communications and Tracking

Maintaining contact between a spacecraft and the ground is a major task. Networks of communications and tracking stations have been set up around the world. They maintain contact with crews, send up commands and computer programs, and receive data from spacecraft. One U.S. system, the Tracking and Data Relay Satellite System (TDRSS), uses satellites. Another employs several radar and radio stations on the ground. Data from these systems is sent to central tracking facilities, such as the one at NASA's Goddard Space Flight Center in Maryland. The Soviet Union relied on a fleet of tracking ships, as does Russia.

This tracking antenna at Goldstone, California, is capable of receiving data from spacecraft at 16,200 bits per second. It transferred TV pictures of all planetary space missions.

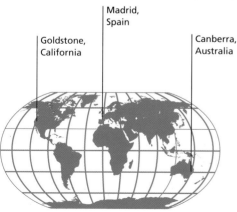

Madrid, Spain

Goldstone, California

Canberra, Australia

Deep Space Network

The Deep Space Network comprises three powerful antennas, each 230 feet (69.9 m) in diameter, positioned around the world at Goldstone, California; Madrid, Spain; and Canberra, Australia. Together they send and receive data from both piloted spacecraft and planetary probes. Information about a spacecraft's precise location, speed, and status is called *telemetry.*

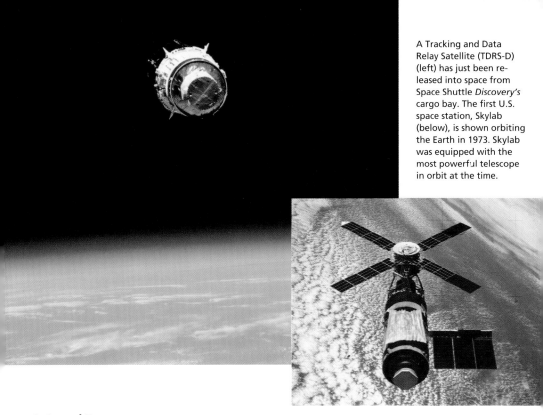

A Tracking and Data Relay Satellite (TDRS-D) (left) has just been released into space from Space Shuttle *Discovery's* cargo bay. The first U.S. space station, Skylab (below), is shown orbiting the Earth in 1973. Skylab was equipped with the most powerful telescope in orbit at the time.

Internal Power

Spacecraft need electrical power for their instruments, for their internal housekeeping systems, and for communicating with Earth. This electrical power comes from batteries, solar cells, or nuclear generators, or a combination of these. Small satellites with limited power needs can operate on batteries. The most commonly used batteries are composed of nickel and cadmium and can last for several years. Newer batteries made of nickel and hydrogen are becoming more widely employed because they are lighter in weight and have a life of ten years or more.

Batteries are also used in conjunction with solar energy systems to provide power for a larger spacecraft when it is not in sunlight. Solar cells, known as photovoltaic cells, are used to convert energy from the Sun into electricity. Some satellites are covered with these cells so that their spin does not affect their power supply. Other spacecraft use large winglike arrays of solar cells to generate electricity. They must maintain a constant attitude in relation to the Sun to keep the panels in sunlight.

Small nuclear generators use the heat given off by radioactive isotopes to generate electricity. Generally, these systems are installed in spacecraft, such as Voyager, that are traveling so far from the Sun and for so long that neither solar energy nor battery power would be adequate.

Astronaut Norman E. Thagard, payload commander, and Canadian payload specialist Roberta L. Bondar (inset) view their surroundings from the aft flight deck of the Space Shuttle *Discovery* in January, 1992. They spent most of their work time in the Microgravity Laboratory (IML-1) science module in the Shuttle's cargo bay. The panoramic view of Earth was taken on an earlier mission of Space Shuttle *Columbia*.

Environmental Control and Crew

No matter what the payload, a spacecraft must maintain the proper internal environment. Both people and instruments need controlled temperature and humidity and protection against harmful radiation.

Providing even the most basic life-support requirements, such as oxygen, water, and food, requires special systems. Supplies have to be carried onboard or delivered to the spacecraft when they run low. Oxygen or a mixture of oxygen and other gases is carried in tanks. Fans keep air circulating through filters to cleanse it of lint and dust and exhaled carbon dioxide.

Water needed for drinking, food preparation, and personal hygiene is also carried in tanks. Filtering systems are provided so the water can be recycled and reused. Skylab was equipped with a shower, but Space Shuttle astronauts take sponge baths.

Food is prepared and packaged on the ground and heated or simply mixed with water in space. In weightlessness, liquids tend to form bubbles that float away, so squeeze bottles contain them. Solid foods that might float around may be eaten from plastic pouches or with thick sauces.

Eliminating body-waste has long been a problem in spacecraft, but the Space Shuttle and space stations make use of specially designed zero-g toilets which separate liquid waste from solid. Urine is expelled from the spacecraft with other waste water. Solid waste is dried and stored for disposal on Earth.

Exercise is important for maintaining the physical fitness of the crew in weightlessness. Psychological needs such as privacy must also be taken into account. And windows are provided so the space travelers can enjoy the view and avoid "cabin fever."

Launching into Space

Once the spacecraft has been designed and built, it must face the challenges of leaving Earth and going into space.

Overcoming Earth's Gravity We are bound to Earth by gravity, and overcoming it requires great speed. The velocity needed to achieve Earth orbit is about 17,500 miles per hour (27,350 km/hr). The velocity needed to send a spacecraft on its way to the Moon or other planets is 7 miles per second (11 km/sec) or about 25,000 miles per hour (40,200 km/hr).

In order for a rocket to liftoff, its thrust must be greater than its weight. The weight of a rocket is mostly propellant, so it takes a great deal of effort to launch even a small payload.

One way to lessen the effect of Earth's gravity is to take off in an easterly direction from a position as close as possible to the equator, adding Earth's own eastward velocity to the rocket's. At the Cape Canaveral launch facility, this adds almost 900 miles per hour (1,450 km/hr) to the rocket's speed.

Staging Another effective way to overcome the problem of escaping Earth is to mount one rocket on top of another. When the first one has burned all its fuel and has accelerated to its top speed, it drops away and the second rocket fires and continues to accelerate. This not only boosts the second rocket to higher speeds, it also lightens the load carried by the second stage. Most spacecraft have three stages (some have four).

Ascent and Acceleration As the rocket lifts off from Earth, Earth's gravitational pull diminishes slowly. At an altitude of 100 miles (161 km), the gravitational pull is only about 5 percent less than it is at Earth's surface. At a distance of about 1,600 miles (2,575 km), it is half as strong. But a rocket's weight and efficiency are affected by another factor. As the rocket burns fuel, it becomes lighter. The effect of these two factors means that a rocket continues to accelerate during the entire time that it is firing.

The Saturn IB space vehicle, carrying the Apollo 7 crew, is photographed after liftoff, more than 35,000 feet (10,640 m) above the Atlantic Ocean. An Airborne Lightweight Optical Tracking System (ALOTS) provides information on launch vehicle performance.

Staging

Each stage of a multi-stage rocket has its own fuel and engine. When the fuel in a stage is burned, that stage falls away from the rocket and the next stage fires, lifting the lighter vehicle higher and increasing its speed. The three-stage Saturn V rocket is used here as an example. (Not all rockets have an escape tower, although the Saturn V does.)

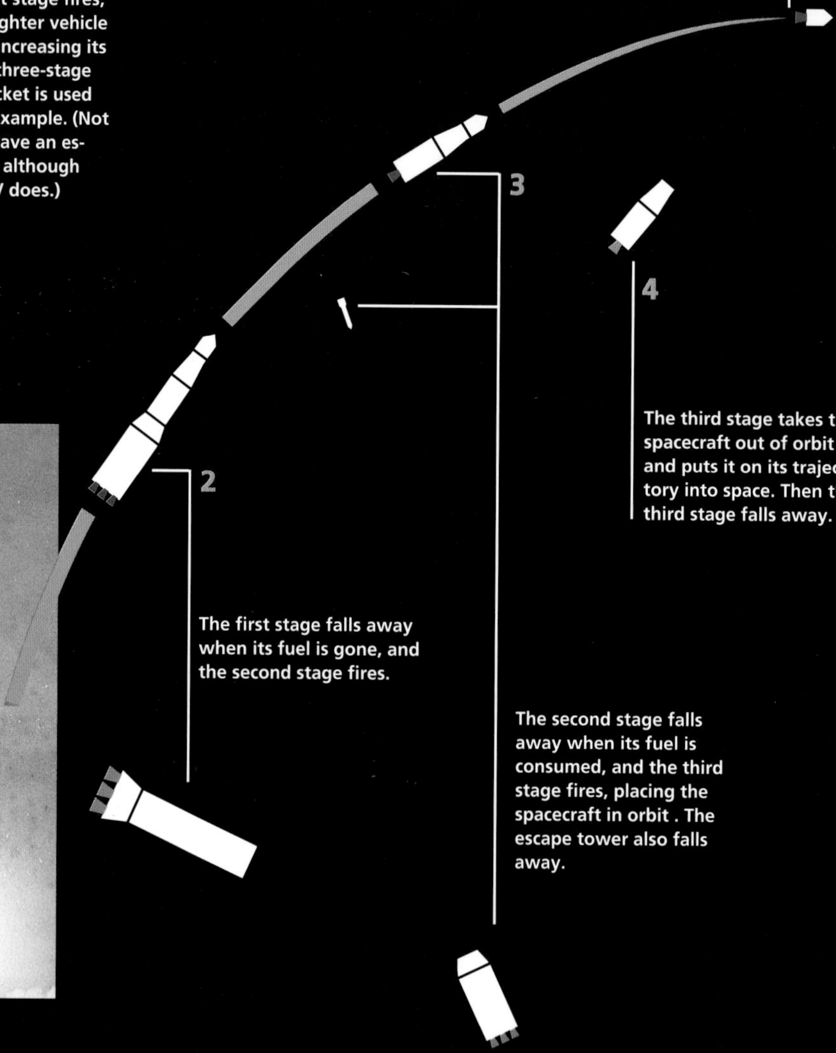

The spacecraft coasts on its flight path under the influence of gravity using its internal propulsion system for acceleration and deceleration.

5

3

4

The third stage takes the spacecraft out of orbit and puts it on its trajectory into space. Then the third stage falls away.

2

The first stage falls away when its fuel is gone, and the second stage fires.

The second stage falls away when its fuel is consumed, and the third stage fires, placing the spacecraft in orbit . The escape tower also falls away.

1

At liftoff, the first stage of the rocket launcher fires. Above it are two additional stages and the spacecraft with its payload.

Flight Trajectories

A trajectory is the path followed by a body moving through the air or space. When you throw a ball, the distance it travels depends on its direction and how hard you throw it. Viewed from the side, the path, or trajectory, of the ball will be a curve. The trajectory of a spacecraft depends on the same principle.

Suborbital Trajectories Rockets in suborbital trajectories, including sounding rockets designed for research in the upper atmosphere, do not go fast enough to go into orbit. After reaching their maximum altitude (when air friction finally slows them down), they fall back to Earth in a long curve.

Orbiting Earth A spacecraft that develops enough speed to balance the gravitational pull of Earth or another body will go into orbit. A spacecraft in orbit is similar to a ball swung on the end of a string, where the inward pull of the string balances the outward centrifugal force of the ball. The speed needed to reach Earth orbit is about 17,500 miles per hour (27,350 km/hr).

If the orbit is high enough to escape the drag caused by the atmosphere of Earth, for example, around 125 miles (201 km) up, the spacecraft can continue in orbit for a very long time. At this distance, it takes about 90 minutes to make one complete revolution. This is known as the spacecraft's orbital period. Farther out, the force of gravity is less, so the speed needed to maintain orbit is lower and the orbital period is longer.

Changing Orbits To move to a higher orbit, a rocket must thrust the spacecraft or satellite forward to accelerate. To reach a lower orbit, a rocket must thrust the spacecraft in reverse to decelerate. While the spacecraft is moving between orbits, it is said to be in a *transfer orbit*. When the proper orbit is reached, another rocket must be fired to correct the speed for the new orbit.

Orbits are elliptical rather than perfectly circular. The closest point of an orbit to Earth is its *perigee*. The most distant point is its *apogee*.

On its first flight test on November 9, 1967, the Apollo-Saturn V spacecraft blasts away from Earth toward lunar orbit.

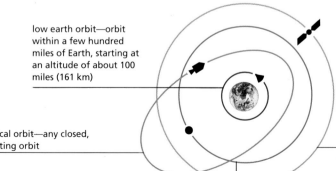

low earth orbit—orbit within a few hundred miles of Earth, starting at an altitude of about 100 miles (161 km)

elliptical orbit—any closed, repeating orbit

higher orbit—orbit 1,000 miles (1,609 km) or more from Earth

Earth Orbits

geosynchronous or geostationary orbit—orbit in which a satellite travels 22,300 miles (35,888 km) above Earth, completes one orbit in 24 hours, and remains over one spot on Earth

Orbits around Earth are often disturbed by forces such as the gravity of the Sun and the Moon, atmospheric friction, or the pressure of solar radiation. These disturbances, which cause a satellite's orbit to change slightly, are called *perturbations.*

Escaping Earth The speed needed to escape the gravitational pull of Earth or another body is called *escape velocity.* The escape velocity for Earth is 7 miles per second (11 km/sec) or about 25,000 miles per hour (40,200 km/hr). A spacecraft that reaches this speed can travel great distances, although its path may be affected by the gravitational pull of the Sun or other bodies in space.

Planetary Trajectories Some spacecraft save fuel and assist their journeys by making use of a *slingshot orbit,* also called a *gravity-assisted trajectory.* In a slingshot orbit, a spacecraft flies close to a planet and uses the gravity of that planet to slow down or speed up. This also changes the spacecraft's trajectory. In effect, the force of the planet's gravity "slings" the spacecraft in a new direction at a new speed.

Launch Timing A spacecraft's path after escaping Earth is planned to take advantage of the changing orbits of the planets. Because the planets and other bodies move in elliptical orbits at different speeds, there are times when they are closer together or farther apart. Favorable launch periods, sometimes called *launch windows,* occur at regular intervals, although these intervals may sometimes be years apart. These are the periods when the flight will require the least amount of fuel and time. Flights can be made at other times, but the velocity must be higher, the fuel expended greater or the payload lighter, and the flight will usually take longer.

Slingshot Orbit

target planets

slingshot trajectory

Sun

Earth

During the weeks before the spacecraft's arrival at its destination, orbital information is calculated and refined. Onboard rockets are used to make any corrections in the spacecraft's path.

Orbiting Other Bodies If a spacecraft reaches the correct speed in relationship to a given moon's or planet's gravity, it will begin to orbit the moon or planet. These trajectories are called *capture trajectories* because the spacecraft is captured by its destination's gravity.

Getting Around in Space

Guidance and Navigation Because of the effects of the motions and the gravities of the various bodies in space, getting from one place to another can be very complicated. Navigation is usually controlled by one or more of three systems.

Most spacecraft rely on *inertial navigation* for guidance. Inertial systems measure the effect of outside forces, such as a change in direction or acceleration, on three spinning gyroscopes. A computer is used to measure and compare changes in the spin of the gyroscopes and give information on the spacecraft's movement. These computers, which are sometimes located on the ground and sometimes onboard, tell the spacecraft which rockets to fire to correct or change its movement.

For spacecraft in orbit around Earth, radio and radar are used to track their position and to order the spacecraft to move. This is known as *radio navigation*.

Celestial navigation uses the stars for guidance. Small electronic telescopes are aimed at several stars. Because the stars appear to be stationary, the onboard computer can calculate the spacecraft's position from their direction.

Rendezvous When two spacecraft approach but do not touch, as in some Gemini and Soyuz missions, it is called *rendezvous*. Because spacecraft are traveling at high speeds in space, bringing two together is a very delicate operation. Onboard radar and computers are used to give precise information about relative positions, distances, and speeds. On U.S. spaceflights, rendezvous has been directed by the human pilots. Soviet spacecraft have relied on automatic systems.

Docking Docking occurs when two spacecraft actually touch and connect. Docking first occurred during a 1966 Gemini mission prior to the Apollo lunar missions. The first international docking took place during the Apollo-Soyuz Test Project in 1975. Docking is an important maneuver for space stations such as Skylab, Salyut, and Mir.

Getting Back

Return At the end of a mission, some spacecraft (chiefly those carrying people) enter a trajectory that will carry them back to Earth, in the return phase of the space mission.

Reentry When the spacecraft returns to Earth's atmosphere it begins the phase of its mission known as reentry. Friction of the atmosphere acts as a brake on the spacecraft, slowing its speed. But this friction also heats the outer skin of the spacecraft to very high temperatures. Early spacecraft carried heat shields which were designed to protect the spacecraft and its contents. Mercury and Apollo space capsules, for example, were equipped with

The Apollo 15 spacecraft splashed down in the Pacific Ocean—on completion of the fourth piloted lunar landing mission—on August 7, 1971.

heat shields that, in part, gradually melted and vaporized under the intense heat. As the material flowed away, it carried the heat with it. Melting heat shields also produced some exciting, fiery displays for the astronauts inside the spacecraft.

The Space Shuttle uses a different method. Small ceramic tiles made of highly heat-resistant silica cover the surface of the spacecraft. These tiles don't burn away, but are designed to withstand the heat directly.

One important factor in reentry is making sure that the spacecraft is properly aligned. If the angle of reentry is too steep, both the heat and the pressure forces are likely to damage the spacecraft or even destroy it. If the angle is too shallow, the spacecraft may actually bounce off the atmosphere and back into space, like a rock skipping off water. Some spacecraft, such as Apollo, use this skipping effect to slow down during reentry. A returning Space Shuttle glides in a series of banked S-curves in order to slow its descent.

Recovery At the end of their flights, many spacecraft deploy parachutes to slow their descent. The Mercury and Gemini capsules and the Apollo Command Modules landed in the ocean, in a landing called a *splashdown,* and were met by helicopters and ships. Soyuz spacecraft make soft landings on land and are met by helicopters and trucks. The Space Shuttle, which lands on a runway like an airplane, uses parachutes to slow its speed after touchdown.

Commander Robert Overmyer and pilot Frederick Gregory land the Space Shuttle *Challenger* (below) at Edward's Air Force Base, California, on May 6, 1985. This Spacelab 3 mission provided valuable data on the microgravity sciences.

After a successful splash-down in the Pacific, Apollo 15 astronauts David R. Scott and James B. Irwin (above) wait their turn to be hoisted by helicopter to the nearby recovery ship where the third crew member, Alfred M. Worden, has just been taken.

From Dreams to Reality

Who would build the machines that could ascend beyond our atmosphere and into space? Early in the twentieth century several brilliant theorists made parallel progress around the world and, in 1926, Robert Goddard launched the first liquid-propellant rocket. The 41-foot flight from a cabbage patch fore-shadowed the 1967 launch of Wernher von Braun's giant Saturn V developed to lift astronauts to the Moon.

Wernher von Braun, early in his career, positions a rocket for an experiment at the German Rocket Society's rocket flight field in Berlin, Germany, about 1930.

In the late 1700s, Indian soldiers used war rockets like these against British colonial troops in India. Iron tubes filled with gunpowder were attached to sword blades or long bamboo sticks.

Early Rockets

The earliest known rocket is the gun powder rocket. This rocket was small, inaccurate, and short-range. For centuries, use of the powder rocket was confined to weaponry, signaling, and fireworks.

Chinese Rocketry Elementary rocket technology was introduced by the Chinese. Most sources believe this innovation dates from around the mid-eleventh to the twelfth century.

Chinese incendiary arrows were coated with an easy-to-burn powder—probably a mixture of charcoal, salt, and sulfur. These were lit and shot from bows to frighten the enemy or used as incendiary devices. These so-called "fire arrows" were easily extinguished, however. By about A.D. 1200 the Chinese had formulated gunpowder by replacing salt with saltpeter (potassium nitrate).

Rockets in Europe and India Although rockets were used experimentally in Europe by the sixteenth century, enthusiasm for military use of rockets did not surface until later, largely as the result of a military campaign in India between 1770 and 1800. Reports reached England that the Indians had used rockets against British troops. These rockets were reported to be larger

Early Rocket Experiments

Research for the nonmilitary use of rockets also began in China. For example, it is reported that around the year A.D. 1500 a Chinese man named Wan Hoo had 47 black powder rockets attached to his sedan chair. They were then ignited at the same time by 47 servants. Wan Hoo was never seen again.

Fireworks and rockets share a common heritage. Here, fireworks are set up in London's Green Park to celebrate the Peace of 1814.

100 Pr Rocket Congreve A.D.1815.

This 32-pound (15 kg) Congreve rocket is typical of the military rockets of the early 1800s. Later, Congreve rockets were used in ship rescues and to send up flares.

than those used for amusement in England. Their weight ranged from 6 to 12 pounds (2–5 kg), and they were stabilized by a 10-foot (3 m) bamboo pole. Their range was estimated to be between a mile and a mile and one-half (1.6–2.4 km). Although these rockets lacked accuracy, the sheer numbers of those used caused a great deal of damage.

Congreve Rockets Several European military men and scientists began to experiment with rockets, but only one was notably successful: the Englishman William Congreve. After personally buying the largest skyrockets available in London, Congreve began to test them in about 1802. He found their range to be between 500 and 600 yards (450–550 m)—less than half the range of the Indian rockets. He obtained permission to use government laboratories and firing ranges and began designing rockets that eventually attained a range of 2,000–3,000 yards (1,800–2,700 m). His later rockets, with an average weight of 30 pounds (14 kg), carried either incendiary or explosive warheads.

Congreve rockets were used extensively during the nineteenth century in Europe and by the colonial powers in Asia, Africa, and the Americas. For example, they were used effectively by the British to firebomb Boulogne in 1806 and Copenhagen in 1807. They also rained fire on the Americans during the British bombardment of Fort McHenry in 1814. That battle of the War of 1812 was immortalized by Francis Scott Key in the words of the U.S. national anthem, ". . . and the rockets' red glare, the bombs bursting in air . . ."

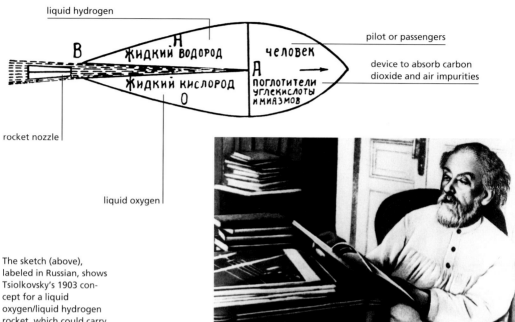

liquid hydrogen

В

ЖИДКИЙ ВОДОРОД

Н

ЖИДКИЙ КИСЛОРОД

О

ЧЕЛОВЕК

А

ПОГЛОТИТЕЛИ УГЛЕКИСЛОТЫ И МИАЗМОВ

pilot or passengers

device to absorb carbon dioxide and air impurities

rocket nozzle

liquid oxygen

The sketch (above), labeled in Russian, shows Tsiolkovsky's 1903 concept for a liquid oxygen/liquid hydrogen rocket, which could carry a pilot or passengers in space. Tsiolkovsky (right) studies at his home in Kaluga, Russia.

Theories and Experiments

Early in the twentieth century, technical curiosity as well as written fantasies of trips to the Moon and elsewhere in space, inspired the work of four great achievers in the field of spaceflight: Konstantin Tsiolkovsky in Russia, Robert Goddard in the United States, Hermann Oberth in Germany, and Robert Esnault-Pelterie in France. These men laid the theoretical and experimental foundations of rocket propulsion that led to stunning accomplishments in spaceflight within the span of only a few decades.

Konstantin Tsiolkovsky

The son of a Russian forester, Konstantin Tsiolkovsky's serious illness as a child left him almost totally deaf and made it difficult for him to attend school. Instead, he studied intensely on his own, particularly mathematics, physics, and astronomy. This self-acquired knowledge enabled him to earn his living as a teacher. It also led to his great interest in space travel.

Tsiolkovsky later recalled that when he was about eight years old, he was given a small balloon—not a common toy at that time. His fascination with it inspired his later work on balloons, including the building of a metallic dirigible which he began in 1885. But Tsiolkovsky realized that balloons could not operate in the vacuum of space, and he began to think about how spaceflight might be accomplished.

> "The Earth is the cradle of humanity,
> but mankind cannot stay in the
> cradle forever."
> —Konstantin Tsiolkovsky

Tsiolkovsky is credited with developing the basic theory of rocket propulsion and with proposing the use of liquid propellant rockets. He presented his theories in an article *The Exploration of Cosmic Space by Means of Reaction Devices*, published in 1903.

Tsiolkovsky concluded that for a spaceship to travel between the planets, in mostly empty space, it would need to have a sealed cabin with oxygen reserves and a system for air purification. He calculated the speed and the amount of fuel that a rocket would need to overcome the force of gravity and go into orbit. He suggested that the greatest velocity of exhaust gases could be obtained with liquid chemical propellants, most likely liquid oxygen and liquid hydrogen. He theorized about the medical effects of zero gravity. Later, he suggested the use of artificial Earth satellites, including occupied platforms, as way stations for interplanetary flight, and he proposed the theory of multistage rockets.

Though he did not receive recognition until late in life, Tsiolkovsky is now acknowledged as a great pioneer of space travel. His work helped stimulate an interest in the use of liquid propellants in rockets in the Soviet Union long before World War II.

The Soviet launch of Sputnik on October 4, 1957, occurred a month after the hundredth anniversary of Tsiolkovsky's birth.

Robert Goddard

Robert Goddard was working on similar matters in the United States without knowledge of the work of Tsiolkovsky or others. But unlike his contemporaries, Goddard didn't stick to pure theory. He verified his theories with experiments and devised hardware to put them into practice. He was the first of the early spaceflight visionaries to combine scientific insight with practical experiments in propulsion.

Robert Goddard poses with one of his rockets in Roswell, New Mexico, in 1935.

In 1919 the Smithsonian Institution published Goddard's paper titled *A Method of Reaching Extreme Altitudes*. The paper contained information about tests Goddard conducted and the conclusions based on those tests. He stated that a rocket could be designed to reach the Moon and that it could explode a load of flash powder to signal its lunar arrival. This idea was picked up by the American press, and Goddard was thereafter known as the "moon rocket man," the odd professor from Massachusetts who believed, as other scientists of the time did not, that a rocket could operate without air, in a vacuum. Goddard was a shy man, and as a result of this public ridicule, he became much more secretive about his work with rockets.

During World War I, Goddard developed a solid-propellant rocket that was successfully demonstrated to military observers in 1918. After the armistice ended the war, however, the military quickly lost interest in this potential weapon.

While majoring in physics in college and graduate school, Goddard had pursued his interest in spaceflight and come to the realization that rocket propulsion was the key to space mobility. He proved in a laboratory experiment that, in spite of popular opinion, rocket thrust was effective in a vacuum. He also soon realized that liquid propellant was the only type of propellant capable of taking a rocket beyond Earth's gravitational field. He theoretically determined that liquid oxygen and liquid hydrogen were the best chemical propellants, though liquid oxygen and gasoline might be more practical.

On March 16, 1926, his rocket fueled with liquid oxygen and gasoline made the first liquid-propellant rocket flight in history, flying 41 feet (12.5 m) up from his Aunt Effie's cabbage patch in Auburn, Massachusetts.

A few days after this 1926 photograph was taken, Robert Goddard (below) successfully launched this liquid-propellant rocket. Goddard did much of his later work in the New Mexico desert, where his team (below right) tested this rocket in September 1931. Goddard stands to the immediate right of the rocket.

From 1930 to 1942, Goddard worked at Roswell, New Mexico, financed by the Guggenheim Fund for the Promotion of Aeronautics. There his rockets achieved altitudes of up to one mile (1.6 km). From 1942 until his death in 1945, Goddard worked for the U.S. Navy at Annapolis, Maryland.

Goddard did not live to see the Apollo 11 mission and men landing on the Moon in 1969. Speaking for her husband, his widow Esther said, "That was his dream, sending a rocket to the Moon. He would just have glowed."

Hermann Oberth

Born in a region of Austria-Hungary that is now in Romania, Hermann Oberth read Jules Verne's *From the Earth to the Moon* and *Around the Moon* at the age of eleven, and from that time on he began thinking about space travel.

Hermann Oberth was not an inventor but a theorist. More than Goddard or Tsiolkovsky, he sparked the rise of modern rocketry in Europe through the German Society for Space Travel, the Verein für Raumschiffahrt (VfR). The mission of the VfR was to spread the idea that the planets were within reach of humanity if humanity were willing to struggle toward that goal.

In 1922 Oberth's Ph.D. dissertation on rocket design was rejected as being too improbable. In 1923 Oberth published *The Rocket into Interplanetary Space*, in which he stated that science and technological knowledge were now sufficiently advanced to construct "machines" that could go beyond the atmosphere and orbit Earth. These machines eventually would be able to remain in space and not "fall back to Earth." They would even be able to go beyond the "zone of terrestrial attraction." He wrote that these machines would fly to the Moon and other planets. Further, they would be able to carry people, "probably without endangering their health." He even examined such problems as space food, space suits, and space walks.

German scientists met in 1930 as the Chemisch-Technische Reichsanstalt (comparable to the U.S. Bureau of Standards) certified the performance of one of Oberth's rocket engines. Oberth stands to the right of the rocket; Wernher von Braun is second from the right.

Oberth's book, while interesting to the general public, was criticized by his peers. Several astronomers stated that space travel was a very nice and interesting idea but was, of course, impossible. And a physician added that as soon as people left Earth's atmosphere, they would be subject to the gravity of the Sun, the power of which would squash their bodies. But, regardless of this criticism, Oberth's theories of spaceflight fired the imagination of an entire generation of young engineers.

Oberth became a German citizen in 1940. During World War II, he witnessed Wernher von Braun's development of the V2 rocket at Peenemünde in Germany. Oberth also worked on solid-propellant antiaircraft rockets during the war.

In 1955 Oberth left Germany and joined Wernher von Braun in the United States, doing some work on the American space program until his retirement and return to Germany in 1958. Long before he died in 1989, Oberth had made a significant contribution to astronautics. He established the theoretical relations that explain the connections between the consumption of fuel, the exhaust gas speed, the launch phase, the duration of the flight, and the distance traveled. From these theories he was able to derive the fundamental laws governing the design of rockets.

U.S. Army rocket researchers, headed by Maj. Gen. Holger Toftoy (standing left), brought a team of German rocket scientists to the United States for postwar research: (left to right) Ernst Stuhlinger, Hermann Oberth, Wernher von Braun, and Robert Lusser.

Robert Esnault-Pelterie

A French pioneer aviator who, as early as 1912, considered aviation to be a step along the road to conquering space, Robert Esnault-Pelterie abandoned his promising career in aviation to devote himself to astronautics after World War I. By 1930 he had published a major work called *L'Astronautique (Astronautics)*. Among his many contributions, he demonstrated the possibility of inertial navigation, guidance of the spacecraft by means of self-contained, automatic devices.

Esnault-Pelterie was convinced that the most significant problem to be solved before putting a person in space was finding the best fuel. Between 1930 and 1940 he searched for an ideal fuel, and, like Oberth, conducted some experiments. But, with the invasion of France in 1940, his work temporarily stopped. After World War II, he continued his work toward the promotion of spaceflight.

Rocket Mania: Trains, Planes, and Automobiles

The rocket-driven sled, Valier Rak Bob 2 (inset), waits for a run on the ice in 1929. The rocket car, Volkhart-R.1, raced up to 40 mph (64 km/hr) in Berlin in 1928.

In the late 1920s and early 1930s, experimenters in the United States, Germany, and Russia tested a variety of rocket engines in cars, gliders, railcars, and sledges—transport vehicles on low runners usually drawn by horses or dogs.

Applying the newly explored principles of rocketry meant more explosions than takeoffs. Excited by the prospect of rocketing into the sky, a few daredevils even belted rocket engines around their waists in futile attempts to launch themselves.

The first rocket plane in history flew on June 28, 1928, in Germany. The next year, Fritz von Opel piloted a glider plane fueled by 16 powder rockets that sustained flight for 75 seconds.

In Berlin, Germany, in 1928 , Max Valier and von Opel joined forces to develop the Rak 2, a car with 24 powder rockets which later sped down a racetrack at 106 miles per hour (170 km/hr). Eight months later, Valier guided a rocket-propelled sledge on a frozen peak in the Bavarian Alps. Valier died in an explosion in 1930 while testing the Rak 7, a car fueled by liquid oxygen and kerosene.

While none of these inventions advanced rocket technology, they certainly reflected a growing fascination with it.

Sergei Korolëv began his experiments in rocketry in the Nakhibino forest near Moscow (above). There, he and members of GIRD (Group for the Study of Jet Propulsion) surround their GIRD-X rocket in 1933. Korolëv's original design for piloted spacecraft became the Soyuz series. Soyuz 19 (right) prepares for liftoff in 1975.

Space Pioneers

The work of Tsiolkovsky, Goddard, Oberth, Esnault-Pelterie, and others set the stage. Significant achievement now depended on the development of multistage rockets of unprecedented power that could take instruments, animals, and finally humans into space. What was needed was an extraordinary engineer, capable of designing and building such rockets. Answering the need were Sergei Korolëv in Russia and Wernher von Braun first in Germany, then in the United States—two exceptionally talented men whose engineering and organizational skills turned concepts and theories into real programs and operational vehicles.

As the first piloted Soviet spacecraft made its successful flight in April 1961, Korolëv, considered the father of the Soviet space program, talked by radio with cosmonaut Yuri Gagarin on board Vostok 1.

Sergei Pavlovich Korolëv

Born in 1907 in what was then Russia and is now Ukraine, Sergei Pavlovich Korolëv built his first glider when he was 18. In 1930 he left flying school to devote all his time to rocket development. Beginning in 1931, he directed the Moscow group known as GIRD (Group for the Study of Jet Propulsion). This organization was both a design office and a factory for building rocket prototypes. Under his direction, the first Soviet liquid-propellant rockets and winged engines were built and tested before World War II. During this time Korolëv wrote his first theoretical, scholarly papers on the possibility of piloted spaceflight.

Under Stalin in the late 1920s and 1930s, the Soviet secret police arrested and imprisoned many intellectuals, scientists, and engineers. In 1938 Korolëv was imprisoned, along with aircraft design engineer Andrei Tupolev and others, and forced to work in a scientific labor camp. When the Germans invaded the Soviet Union during World War II, the camp was moved from the Moscow area to a location east of the Ural Mountains, out of reach of the German armies.

After the war, Korolëv returned to rocket engine development. This time, he was assigned to coordinate the assimilation of German rocketry expertise. As chief designer of guided missile development, Korolëv was able to organize a design bureau that produced the first successful Soviet intercontinental ballistic missile in August 1957 and launched the world's first satellite less than two months later. The international propaganda success of the October 4, 1957, launch of Sputnik allowed Korolëv to pursue his two-decade-old theories about launching a human into space. Soviet leader Nikita Khrushchev recognized that the ability to launch satellites, probes, and later people into space supported his political and military challenge to the United States. The Space Race was born out of the American–Soviet Cold War competition for world leadership.

In less than a decade after the launch of Sputnik, Korolëv was able to organize and manage a series of space "firsts" that fueled the Space Race through the 1960s. During that period, the Soviets launched into space the first man, Yuri Gagarin (April 12, 1961) and the first woman, Valentina Tereshkova (June 16, 1963). They also accomplished the first flight carrying more than one person, Voskhod (October 12, 1964), and the first space-walk, performed by Aleksei Leonov (March 18, 1965).

The Soviet Union also successfully crash-landed a probe on the Moon and photographed the far side of the Moon (Luna 2 and Luna 3, 1959), and attempted to send probes to Mars and Venus. By the mid-1960s, Korolëv undertook a new mission—to design a piloted spacecraft that would reach the Moon. His preliminary design, the Soyuz spacecraft, is used today as a piloted spacecraft that ferries cosmonauts to the space station Mir.

Called the father of the Soviet space program, Korolëv's contribution to the development of space exploration was enormous. His identity was a carefully guarded state secret, and he was officially referred to in the press and other media only as the "Chief Designer of Launch Vehicles and Spacecraft." It was only after his death, during routine surgery in 1966, that the Chief Designer was formally identified by name.

Wernher von Braun

Wernher von Braun, who was born in Germany in 1912, showed an early inclination toward science and music. Having met Hermann Oberth and other pioneers in the field, he became enamored with the idea of space travel and rocketry in the mid-1920s, and became an enthusiastic member of the VfR. Rocketry was an endeavor that would occupy him for the next 50 years and earn him worldwide fame.

In 1932 von Braun took charge of research and development of rockets as military weapons for the German army at a small testing facility near Berlin. The first rocket developed by von Braun and his group was the A1, soon redesigned as the A2.

In August 1969, shortly after the first Moon landing, Wernher von Braun stands proudly with an Apollo capsule and a Saturn V rocket.

Larger, more sophisticated military testing facilities were needed and these were built at the village of Peenemünde in northeastern Germany. As the technical director of this facility, von Braun embarked on the development of an operational military missile—the A4. This rocket was designated by the German Propaganda Ministry

The V2 (Vengeance) Rocket

By its third launch in Germany on October 3, 1942, the V2 was a technological success. Weighing about 29,000 pounds (13,000 kg) with a payload of about 2,000 pounds (900 kg) of high explosives, the V2's typical altitude was about 50 miles (80 km) and its horizontal range was about 186 miles (300 km).

The V2's capabilities signaled a new era in both the military and space applications of rockets. The V2 terror weapon of World War II was the clear forerunner of all later rockets and ballistic missiles.

A German V2 rocket (below), one of the most terrifying weapons of World War II, waits on its launching ramp.

as the V2, or the Vengeance Weapon 2. Thousands of V2s were built at a production facility near Nordhausen by concentration camp prisoners. The V2 was a terror weapon responsible for civilian deaths and destruction during World War II.

But von Braun's primary interest was still rockets for space travel. In fact, his arrest along with two other engineers in 1944, was politically motivated and based partly on this preference. The Nazi Gestapo accused the three of saying that their real passion was space travel, not weapons. It followed, then, that they had not applied all their energies to the missile, which constituted sabotage. Von Braun was imprisoned for only two weeks before being freed because of his unique value to the war effort.

In 1945, facing the advance of the Soviet army from the east and the Americans and British from the west, von Braun surrendered to the U.S. Army. He and about 120 of his engineers signed contracts to work with the Army. Soon von Braun's team, the recent enemy, was using captured V2s to teach Americans about rocketry and doing research and test flights at Fort Bliss, Texas, and White Sands, New Mexico. Several years later, they began work on a long-range ballistic missile at Redstone Arsenal in Huntsville, Alabama. In 1956 they launched a three-stage Jupiter C rocket which reached an altitude of 680 miles (1,094 km) and splashed down 3,300 miles (5,300 km) away.

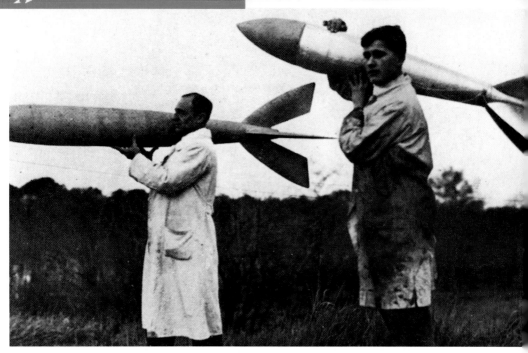

Long before he became technical director of the German army missile program, Wernher von Braun was working on military rockets. Here, in about 1932, a youthful von Braun (right) carries a streamlined Repulsor rocket across a rocket test field. The older scientist is Rudolf Nebel, who worked with Oberth.

It wasn't long before von Braun was able to turn his thoughts back to the idea of space exploration. In 1957, when America faced the shattering realization that the Soviets had successfully launched Sputnik, Secretary of Defense Neil McElroy approved von Braun's preparation of a satellite and rocket for launch by March 1958. In fact, by late January a Juno 1 rocket (a modified Jupiter C rocket with a solid-propellant fourth stage carrying a small satellite) was ready for launch. On January 31, 1958, telemetry signals from the first U.S. satellite, Explorer 1, confirmed that the Juno 1 launch from Cape Canaveral, Florida, had indeed achieved Earth orbit.

The National Aeronautics and Space Administration (NASA) was established on October 1, 1958. On July 1, 1960, von Braun's group became the core of the George C. Marshall Space Flight Center—NASA's major center for the development of launch vehicles and propulsion systems—in Huntsville, Alabama. There von Braun directed the development of rocket launch vehicles including Saturn I, IB, and V.

The engineering and managerial complexities involved in the development and successful launches of the Saturn rockets testify to von Braun's capabilities. Remarkably efficient, each of these rockets was launched on schedule and met performance expectations as well as safety requirements.

In 1961, President John F. Kennedy announced the U.S. goal of "landing a man on the Moon and returning him safely to the earth" by the end of the decade. Von Braun's expertise and the de-

The huge first-stage engines of a Saturn V launch vehicle (above) dwarf their developer, Wernher von Braun. A Saturn V rocket (left) launches the historic lunar landing mission of Apollo 11, which began on July 16, 1969.

velopment of the Saturn V rocket became integral to that endeavor. The first successful launch of the three-stage Saturn V was on November 9, 1967, from Cape Canaveral. After Apollo 7 was launched by a Saturn IB, Saturn V rockets launched all piloted Apollo missions. Thus, when the crew of Apollo 11 became the first humans to land on the Moon in July 1969, von Braun was the key figure behind this amazing achievement.

Research on the Saturn rockets was the focus of von Braun's career with NASA. He remained director of the Marshall Space Flight Center until 1970 when he was transferred to the NASA planning office in Washington, D.C. In 1972, he retired from NASA to enter private industry and focus on other efforts to promote spaceflight. In December 1976 von Braun retired. Before his death in 1977, he was awarded the National Medal of Science by President Gerald Ford, one of many honors bestowed upon von Braun during his distinguished career.

Wernher von Braun played a singularly prominent role in the history of rocketry and spaceflight, seizing opportunities with both the German regime and the United States to pursue these interests. His development of the V2 and the Saturn V altered world history. He was the guiding force behind the world's first ballistic missile, the launch of the first U.S. satellite and first U.S. astronauts, and the development of the most complex rocket ever built, which sent the first humans to the Moon. Much of the success of the U.S. space program can be directly attributed to von Braun's technical genius and managerial competence.

Rocket Launch Vehicles

The countdown ends and massive engines fire, shaking the ground and creating billowing clouds of exhaust. A towering launch vehicle carrying a spacecraft bound for the outer planets roars off the launchpad and leaves Earth behind.

Launch vehicles, the workhorses of spaceflight, deliver human and satellite payloads into space. Although the United States and the Soviet Union led the development of launch vehicles, several other nations now have their own.

The Space Shuttle
Columbia lifts off from
Cape Canaveral on
March 22, 1982.

Launch Vehicles

The launch vehicle is the rocket system that lifts a spacecraft off Earth and gives it enough speed to achieve orbit. Some launch vehicles consist of several successively smaller rockets stacked on top of one another, called *stages*. As each stage exhausts its fuel, it is detached from the vehicle and falls back to Earth as the spacecraft proceeds into space.

Since 1958, when the failure rate of both Soviet and U.S. launch vehicles was nearly 50 percent, the technology, reliability, and performance of launch vehicles have dramatically improved. Many other nations have also developed their own launch vehicles. Launch vehicles can now put spacecraft carrying huge scientific and piloted payloads into orbit.

U.S. Launch Vehicles

Atlas The Atlas family of launch vehicles was named after the mighty god of Greek mythology who supported the world on his shoulders. The Atlas was adapted from the U.S. Air Force Atlas intercontinental ballistic missile (ICBM).

The Atlas family of launch vehicles has taken part in many important projects. The Atlas IIAS successfully launches the AT&T Telstar satellite in December 1993 (above). In 1962 an early version of Atlas prepares to thrust a Mercury capsule into orbit (right).

The original Atlas liquid-propellant launch vehicle was 90 feet (27 m) long and 12 feet (3.7 m) in diameter. Its first design was called a "stage-and-a-half" because it had a single main engine with two side-by-side booster engines, all powered by liquid oxygen and kerosene. All three engines fired at launch, but the two boosters dropped away during the first-stage burn. This Atlas had a takeoff thrust of 350,000 pounds (160,000 kg).

The most significant mission by an Atlas launch vehicle was part of Project Mercury, the primary goal of which was to put a human in Earth orbit, study his or her physical and mental reactions, and bring him or her safely back. This mission was accomplished on February 20, 1962, when an Atlas launched *Friendship 7* carrying John Glenn, the first American to orbit Earth.

A Voice from Space

In December 1958 an Atlas rocket launched the mission designated Project Score which carried a prerecorded Christmas message from President Dwight D. Eisenhower to be transmitted to the people of the world. This message—the first human voice heard from space—was meant to convey the neutrality of space frontiers and the hope that the exploration of space would be achieved with a spirit of cooperation.

Atlas was also combined with two upper-stages, the Agena and the Centaur. Atlas-Agena, last used in 1967, launched space probes such as Mariner and Ranger. The Atlas-Centaur, in current use, is a more powerful version. It has launched many spacecraft, including Surveyor, Mariner, and Pioneer, and numerous military and communications satellites. The Atlas IIAS, the most powerful of the Atlas-Centaur family, can send 8,150 pounds (3,697 kg) into geosynchronous orbit and 19,050 pounds (8,641 kg) into low Earth orbit.

Delta The first liquid-propellant launch vehicle of the Delta family was based on the Thor intermediate-range ballistic missile. From 1960 to 1982 Delta underwent a series of improvements, and eventually 34 different versions were developed. The current version, the Delta II, is completely different in appearance from the Thor-Delta of 1960. The Delta II has a larger first stage, with a more powerful engine and solid rocket boosters strapped to its sides. A solid-propellant third stage can be added to the first two liquid-propellant stages if the mission requires it.

The success rate of the Delta launch vehicle in more than 200 launches has been exceptional, and includes the following launch "firsts": Echo I in 1960, the first communications satellite in Earth orbit to relay voice and TV signals from one ground station to another; Syncom I in 1963, the first geostationary satellite; and Early Bird in 1965, the first Intelsat (International Telecommunications Satellite Organization) satellite. Delta also launched many Explorer satellites, Pioneer interplanetary probes, and most of the satellites in the TIROS and Landsat series.

The current Delta model is 116 feet (35 m) long and 8 feet (2.4 m) in diameter. Delta's first stage, with strap-on boosters, develops a combined thrust of about a million pounds (450,000 kg).

A Thor-Delta prepares to launch the communications satellite Telstar II in 1963. The Telstar satellites pioneered in transmitting television pictures between the United States and Europe.

Titan Titan's first launch was in 1959, and Titan liquid-propellant launch vehicles are still in use today. Like Atlas and Delta, Titan comprises a family of rockets. Titan technology began with the Titan I, a military ICBM propelled by oxygen and kerosene. It was later adapted by NASA for civilian uses.

Titan II, no longer in use, launched the two-person Gemini missions in the 1960s.

Titan IIIEs launched the interplanetary Viking and Voyager missions which require high-velocity escape trajectories. Images transmitted from these missions allowed an awestruck world to view Mars and the outer planets in a degree of detail never before thought possible.

The upgraded Titan IV is the largest, most powerful U.S. expendable launch vehicle currently in use. It is capable of placing approximately 48,750 pounds (22,000 kg) in low Earth orbit, or sending about 12,500 pounds (5,670 kg) of payload into geosynchronous orbit. Titan IV is used to launch large military payloads as well as NASA's unpiloted deep space missions.

Saturn Beginning in 1958 NASA began production of a family of three large liquid-propellant launch vehicles. These were Saturn I, IB, and V. The Apollo lunar missions used the Saturn V launch vehicle.

Saturn I was developed for unpiloted flight tests of the Apollo module. Assembled from components made for other rockets, Saturn I was a two-stage rocket, the first stage consisting of eight clustered engines. The second stage burned liquid hydrogen with liquid oxygen, and had six engines with a thrust of 15,000 pounds (6,800 kg) each.

Titan, like Atlas, is a name from Greek mythology—the Titans were a race of giants. Here a powerful Titan IV with a Centaur upper stage stands ready to launch a 10,000-pound (4,536 kg) MILSTAR communications satellite into orbit from Cape Canaveral on February 7, 1994.

Saturn V: To the Moon

The first stage of Saturn V, weighing nearly 5 million pounds (2.3 million kg), lifted each of the piloted Apollo spacecraft about 41 miles (66 km) above Earth. The second stage then accelerated the spacecraft to a height of 120 miles (193 km), and was jettisoned—at which point the spacecraft was traveling at about 17,000 miles per hour (27,350 km/hr) and had achieved Earth orbit. Finally, the third stage lifted the craft to an altitude of about 190 miles (300 km) and a speed of 24,300 miles per hour (39,110 km/hr), which sent it on a trajectory to the Moon.

In January 1964 a Saturn I placed its payload into Earth orbit. The next two flights, that same year, put development versions of the Apollo capsule into orbit. Its final flights in 1965 launched Pegasus meteoroid detection satellites.

The Saturn IB was a revised version of the Saturn I, with an improved first stage and a new second stage propelled by a single engine. It could place 40,000 pounds (18,160 kg) into low Earth orbit. Its first unpiloted launch was in February 1966. On October 11, 1968, a Saturn IB launched the first three-person Apollo mission, Apollo 7. During 1973 and 1974 Saturn IBs delivered three crews to Skylab. In 1975 three U.S. astronauts were launched by a Saturn IB on the Apollo-Soyuz mission.

Development of the giant, three-stage Saturn V began with one goal: to send humans to the Moon. From 1968 to 1972 its consistent performance and reliability enabled 24 U.S. astronauts to orbit the Moon and 12 of them to walk on it.

The massive Saturn V launch vehicle stood 363 feet (111 m) high. Its first flight took place on November 9, 1967, when it placed 126 tons (114 MT) into orbit. It thundered upward from the launch pad with a thrust of 7.5 million pounds (3.4 million kg). In a mere 2 minutes and 40 seconds its five first stage engines burned 4.6 million pounds (2 million kg) of propellants. One witness remarked after the deafening, earthshaking launch that the question wasn't whether the Saturn had risen but whether Florida had sunk.

Saturn V launched all of the piloted Apollo flights, from Apollo 8 on, including the Apollo 11 mission that accomplished the first walk on the Moon, on July 20, 1969. The final launch of Saturn V was on May 14, 1973, when the Skylab space station (itself a modified third stage of a Saturn V) was put into orbit.

The Saturn V was the largest, most powerful rocket ever launched. In 1969, Saturn/Apollo 11 (left) waits on its mobile launch platform for the start of the first lunar landing mission. A Saturn V (right) launches the Skylab space station on May 14, 1973.

the Vehicle Assembly Building (VAB) where two Space Shuttles can be assembled at once

the Vehicle Assembly Building viewed from a launchpad

the road for the Crawler which moves assembled launch vehicles from the VAB to the launchpad

the Launch Control Center (LCC)

The Kennedy Space Center at Cape Canaveral, Florida, covers nearly 140,000 acres of land and water.

Each launch site is a massive social, technological, and economic enterprise, where a total community dedicated to spaceflight lives and functions. Each site must be able to accommodate the major construction involved, including a technical center, a control center, and the launch complex itself.

Locations for sites are selected for geographic and political reasons. Proximity to the equator increases the efficiency of launches because Earth's rotation enhances rocket launches to the east (for geostationary orbit) and to the north (for near-polar orbits). This means that more weight and therefore more fuel can be carried from a launch site closer to the equator. Most nations, however, do not have this geographic option. The San Marco and Kourou sites are examples of those that are well located to take advantage of this rotational boost.

For safety, to avoid launching over populated areas, locations chosen for launch sites often border oceans. The U.S. launch sites at Cape Canaveral, Vandenberg, and Wallops Island, are examples of such coastal locations.

the Space Shuttle landing runway

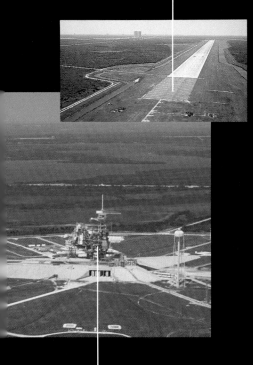

one of two operational Space Shuttle launchpads

Extremes of temperature or harsh conditions such as frequent storms make an area less suitable for launching. For example, seasonal storms pose a threat at Cape Canaveral, and corrosive effects of severe sandstorms at Baikonur in Kazakhstan necessitate transporting launch vehicles in a horizontal position. On the positive side, the sun at Baikonur shines for more than 300 days a year.

Knowledge about existing and planned launch sites scattered across the world is generally available today. In contrast, much secrecy surrounded these locations and facilities during the Cold War. For example, the size and location of some of the Russian launch sites, or cosmodromes, were only revealed to the United States via images from spy planes and data provided by Earth observation satellites. In 1966 the existence of Plesetsk, the most active and northernmost Russian cosmodrome, was announced because of the investigation of a class of British students. They calculated the orbit of the Cosmos 112 flight and concluded that it must have been launched from a new Soviet site—eventually revealed as Plesetsk.

Scout The Scout launch vehicle, used from 1960 to 1994, was a lightweight, solid-propellant booster—the smallest of the basic launch vehicles. Scout's development was a joint effort by NASA and the U.S. Department of Defense to construct a relatively inexpensive rocket from existing solid-propellant motors.

In 1960 it was developed as a three-stage rocket. Later an additional stage was added. Scout's original height was 72 feet (22 m), and it had a liftoff thrust of 104,500 pounds (47,400 kg). It was especially suited for high-altitude probes, small-satellite launches, and reentry experiments.

Pegasus Pegasus, named after the winged horse of mythology, is a three-stage solid-propellant winged rocket. It is one of several new commercially developed launch vehicles for small to moderate payloads. Pegasus is taken to a high altitude by a jet aircraft, then released. The solid-propellant engine of the first stage produces 109,000 (49,500 kg) pounds of thrust. The second stage provides 27,600 (12,530 kg) pounds of thrust. The third stage, at 9,800 (4,450 kg) pounds of thrust, propels the satellites into orbit.

Pegasus replaced Scout as the basic lightweight launch vehicle because Pegasus launch locations and times are more flexible. Thus, bad weather can be avoided and, with no permanent site or ground crew, it is possible to launch more economically.

The initial Pegasus launch took place in 1990, when the rocket orbited a test vehicle called Pegsat. A second flight, in 1991, carried seven small satellites. Flights are planned through the 1990s.

The winged Orbiter shown at the center of the Space Shuttle (far left and right) includes a payload bay large enough to hold a bus. It is used to carry satellites and spacecraft into space. The astronauts live and work in a much smaller crew cabin.

The Space Shuttle The U.S. Space Shuttle, officially called the Space Transportation System, is the world's first reusable launch vehicle, able to make multiple launches rather than being discarded in each mission.

The 184-foot (56 m) Space Shuttle includes the Orbiter, a completely recoverable winged space plane with three liquid-propellant main engines; the External Tank (ET), used only once, which contains liquid hydrogen and liquid oxygen used by the main engines during launch; and twin Solid Rocket Boosters (SRBs), the largest solid-propellant motors ever flown, and the first designed for reuse. Each SRB is 149 feet (45 m) long and holds 1.1 million pounds (500,000 kg) of solid propellant. These boosters, along with the main engines, are ignited at liftoff and create about 3.3 million pounds (1.5 million kg) total thrust. The boosters burn for a little over two minutes, then fall off and parachute into the ocean, where they are recovered for reuse on another mission.

At the back of the Orbiter are three Space Shuttle Main Engines (SSMEs) and two Orbital Maneuvering System (OMS) engines. With a combined thrust of almost a million pounds, the main engines are the most complex, powerful cryogenic liquid-propellant engines ever built.

The External Tank is 154 feet (47 m) long and 28 feet (8.5 m) in diameter. It holds the 633 tons (574 MT) of liquid hydrogen (the fuel) and liquid oxygen (the oxidizer) for the Orbiter's main engines. This is the only major part of the Space Shuttle that cannot be reused. The fuel in the aluminum tank is depleted after only eight and one-half minutes of flight, after which the tank is jettisoned into the ocean. At that point, the two OMS engines place the Orbiter into orbit—and later take it out of orbit for reentry. The Shuttle's Reaction Control System (RCS) consists of 38 engines and 6 smaller *vernier* engines—used to adjust the trajectory, attitude and velocity of the Space Shuttle while it is in orbit.

The Shuttle Heat Shield

Between the Shuttle astronauts and certain incineration during reentry to Earth's atmosphere lie some 28,000 protective tiles, which form a unique jigsaw pattern on the surface of the Orbiter. These tiles, made of silicate fibers with a ceramic coating, can absorb heat as high as 2,700° F (1,482° C), yet throw off heat so quickly that they can be handled without injury immediately after reaching these astronomically high temperatures. Each tile has been individually shaped to fit the contour of the Shuttle's fuselage skin.

The Orbiter returns to Earth by landing on a runway, like an airplane. When it must be transported to another location, it is flown there perched atop a Boeing 747.

The Space Shuttle has been used to launch space probes such as Galileo, to the planet Jupiter, and Ulysses, into orbit around the Sun. The 18th mission of the Space Shuttle, in June 1985, placed four satellites, including U.S., Mexican, and Arab communications satellites, into Earth orbit. It also carried a set of student science experiments. In April 1990 the Hubble Space Telescope was launched by the Shuttle.

Upper Stages of Launch Vehicles

Upper stages are added to increase the payload capability of launch vehicles for more complex space missions. Upper stages, several of which are described below, are fired after the payload is released from its primary launch vehicle.

Agena The oldest U.S. upper stage, the liquid-propellant Agena has met with great success, including its use as the first space-docking target. On that mission, on March 16, 1965, the crew of Gemini VIII rendezvoused and docked with a pre-launched Agena stage. Agenas also launched the Mariner probes of Mars and Venus as well as several military spy satellites. Agenas have been used as the upper stage on Atlas, Thor, and Titan IIIB launch vehicles.

Centaur The primary purpose of this liquid-propellant upper stage, named for the legendary creature that was half man and half horse, has been to launch spacecraft on planetary missions.

The Inertial Upper Stage (IUS) is used to deploy the Tracking and Data Relay Satellite from the Space Shuttle *Discovery* on September 29, 1988.

When NASA was formed in 1958, developing the Centaur was one of its first goals. Used from 1963 onward, it was the first upper stage to use liquid hydrogen and liquid oxygen.

The upper stage for Titan III is a Centaur. The most powerful Atlas variant is the Atlas-Centaur. Versions of Atlas-Centaur have sent Surveyor probes to the Moon, Mariner missions to Mars and Mercury, and placed large communications satellites in Earth orbit.

Because solid-rocket motors are safer than liquid-propellant engines, NASA canceled use of the Centaur for Space Shuttle missions and replaced it with the Inertial Upper Stage.

Inertial Upper Stage (IUS) The IUS is designed to launch heavy payloads from both Titan and the Space Shuttle. It can place a spacecraft in a geosynchronous orbit or on the appropriate trajectory for an interplanetary mission. The IUS was used for the Magellan mission to Venus, the Galileo mission to Jupiter, and the Ulysses mission to the Sun via Jupiter.

The IUS is a two-stage vehicle with a combined thrust of more than 60,000 pounds (27,220 kg). Each stage is a solid-rocket motor. It is 17 feet long (5 m) and 9.5 feet (2.5 m) in diameter and weighs about 32,500 pounds (14,750 kg).

Transfer Orbit Stage (TOS) TOS is a liquid-propellant upper stage that can take payloads weighing 6,000 to 13,000 pounds (2,700–5,900 kg) from Space Shuttle orbit to a geosynchronous transfer orbit. It is also used as an upper stage for Titan. TOS measures 11 feet (3.4 m) long and 11 feet (3.4 m) in diameter and weighs about 24,000 pounds (10,890 kg) when fueled.

Soviet Launch Vehicles

Vostok Standing 126 feet (38 m) tall, with one main rocket engine and four boosters, Vostok ("east"), the first operational Soviet ICBM, was designed by Sergei Korolëv in the late 1950s. Vostok launched early, unpiloted Luna, Electron, and Meteor satellites as well as piloted Vostok spacecraft.

Vostok spacecraft launched by Vostok rockets completed several historic missions. Vostok 1, launched on April 12, 1961, was the world's first piloted spacecraft; cosmonaut Yuri Gagarin made a single orbit around Earth and returned safely. Vostok 2, launched the same year, put a cosmonaut into orbit for 17 revolutions of Earth. By 1963 the piloted flight of Vostok 5 had completed the longest one-person flight to date—nearly five days—and Valentina Tereshkova, the first woman in space, had completed a three-day orbital mission aboard Vostok 6.

Soyuz Soyuz ("union") was derived from the Semiorka rocket, designed by Korolëv, and consists of four first-stage boosters attached to a central body. Each booster has a rocket engine, with four combustion chambers and four nozzles, which burns kerosene and liquid oxygen. The first stage, taken from the intercontinental SS-6 Sapwood missile developed by Korolëv, launched Sputnik in 1957, and later Sputniks 2 and 3. Upper stages were added to take crews to orbiting space stations. The second stage has two engines.

Soyuz is the basic workhorse for missions to low Earth orbit. It is 163 feet (50 m) high and can place 7.5 tons (6.8 MT) into low orbit. It has been used for launches of piloted Soyuz spacecraft and Progress supply vehicles. Soyuz rockets were used in the Voskhod program and are still in use today to send Soyuz spacecraft to the Mir space station.

The Soyuz T-II spacecraft (above) blasts into orbit in the spring of 1980. Much of Korolëv's original design for the Vostok rocket has remained in the Soyuz.

The Proton launch vehicle, seen during assembly (left), sent the Soviet spacecraft Luna 15 to the Moon and also launched several Venera missions to Venus. In 1984 Proton launched Vega (above), which entered orbit around Venus in June 1985 and went on to meet Halley's Comet in 1986.

Cosmos The primary use of the Cosmos series has been for launching military, scientific, and communications satellites. Based on a military missile, the original Cosmos was a small two-stage launch vehicle. Beginning in March 1962, it launched more than one hundred scientific satellites in the Cosmos series. It was phased out in 1977.

The Cosmos presently in use is a small vehicle about 105 feet (32 m) in length. It is based on the Skean medium-range ballistic missile, to which a restartable second stage has been added. Used to launch hundreds of small satellites, Cosmos generates 176 tons (160 MT) of thrust in the first stage.

Proton Versions of the Proton rocket have launched heavy spacecraft designed for lunar and planetary exploration. Proton has a six-engine first stage, to which a second stage and escape stages can be added. It can place about 40,000 pounds (18,150 kg) into low Earth orbit, send 11,400 pounds (5,170 kg) to the Moon, or propel 10,000 pounds (4,540 kg) as far as the inner planets.

Proton rockets attempted to launch unpiloted Cosmos spacecraft to the Moon beginning in 1967, but not until July 21, 1969, did the unpiloted Luna 15 reach the Moon's surface—one day after Apollo 11's historic lunar landing. The Proton never carried cosmonauts because the Soyuz was adequate for that purpose. In April 1971 the Proton launched the first Salyut ("salute") space station. For ten years it has been used to launch unpiloted spacecraft to Venus; and it launched the present space station Mir ("peace" or "world") in February 1986.

Energiia Used only twice to date, Energiia was first launched in 1987. With two recoverable stages and a launch mass of about 2,400 tons (2,180 MT), it reportedly can place a payload of about 100 tons (91 MT) into low Earth orbit, and about 18 tons (16.3 MT) into geostationary orbit. Energiia is powerful enough to launch an estimated 32 tons (29 MT) into lunar orbit, and about 28 tons (25 MT) to Mars. A version of this launch vehicle was used for the unpiloted test flight of the Soviet space shuttle *Buran* in November 1988.

The Energiia is 202 feet (61 m) tall and weighs 4.4 million pounds (2 million kg). Its first stage generates 6.6 million pounds (3 million kg) of thrust. Its core tank has four liquid-hydrogen/liquid-oxygen engines. The four reusable boosters are fueled with kerosene and liquid oxygen.

The Energiia launch vehicle is joined with the space shuttle *Buran* on the launchpad, prepared for liftoff. Energiia has been closely associated with this shuttle since *Buran* has no launch system of its own.

Ariane 4 lifts off for its first flight in June 1988 from "Europe's space-port," ESA's launch site at Kourou, French Guiana (on the northeast coast of South America).

European Launchers

Diamant To develop its first civilian satellite launch vehicle, the Centre National d'Études Spatiales (National Center of Space Studies), or CNES, continued work begun by the French military. The basis for the liquid-propellant Diamant ("diamond") project was experimental rocket motors used in missiles and research rockets. These smaller upper atmosphere research rockets were also named for gems: Agate, Topaze, Rubis, Emeraude, and Saphir. Diamant drew most from the two-stage Saphir rocket, to which a third stage was added. The first stage single engine was propelled by liquid fuel; the second and third stages of Diamant used solid propellant.

Three different Diamant rockets were constructed, each with a capacity greater than that of the previous one. The initial launch was in November 1965, when an A-1 satellite was successfully orbited. Of the eight additional Diamant launches, six were successful. The final launch took place in September 1975.

Ariane France proposed, and provided about 60 percent of the funding for, Ariane, the European launch vehicle selected in 1973 by the European Space Agency (ESA) for development. The Ariane family of liquid-propellant rockets was developed to provide a capability to launch European scientific and communications satellites and to establish a commercial launch service for other nations' use.

Ariane was first launched successfully in 1979, and the eleventh and final Ariane 1 launched the ESA Giotto probe to Halley's Comet in 1985. In 1983 the Ariane 2 rocket was introduced. On its final mission in 1989, Ariane 2 launched the Swedish Tele-X satellite.

Ariane 3 was similar to Ariane 2, but added two solid-propellant strap-on boosters. The second stage had a single engine similar to that of the first stage, and the third stage used cryogenic propellants. With a total thrust of about 850,000 pounds (385,500 kg), Ariane 3 could put 5,700 pounds (2,585 kg) into geosynchronous transfer orbit.

Ariane 4, intended for use through the 1990s, was first launched in 1988. A variety of rocket combinations—using either solid or liquid propellants—can be fitted to its first stage which is longer than its predecessor's. A thrust of 150,000 pounds (68,100 kg) is generated by each engine. The third stage is an improved version of Ariane 3.

The latest vehicle in the Ariane program is the Ariane 5, designed to be more cost effective and reliable. Ariane 5 was to be used as the booster for a European space plane called Hermes until that program was cancelled in 1992. The first scheduled launch of Ariane 5 is in 1996.

Black Arrow This lightweight British launch vehicle had as its ancestor the liquid-propellant Black Knight ballistic missile. Four Black Arrows were built and launched between June 1969 and October 1971, but in 1973 the British government abandoned the program. The first stage contained 13 tons (11.8 MT) of liquid propellant (hydrogen peroxide and kerosene), and the second stage 3 tons (2.7 MT). The third stage was a solid-propellant rocket.

Launch Vehicles of Other Nations

China's Long March In 1970 the People's Republic of China first used its own launch vehicle. Since then it has reportedly launched two to three missions a year, primarily unpiloted satellites. Some contain experiments performed in weightlessness, and others are reconnaissance satellites.

Chang Zheng-1 ("Long March 1") was developed with the help of the then USSR and launched in April 1970. It weighed 82 tons (74 MT) at liftoff and had two liquid-propellant stages and a solid-propellant upper stage. It stood 97 feet (30 m) tall and could place 660 pounds (300 kg) into low Earth orbit.

The European Space Agency

In 1975, the European Launcher Development Organization and the European Space Research Organization merged into the European Space Agency (ESA). At present, ESA member states include Austria, Belgium, Denmark, Finland, France, Germany, Ireland, Italy, the Netherlands, Norway, Spain, Sweden, Switzerland, and the United Kingdom.

Successive improvements on Long March 1 have increased the vehicle's capabilities. Long March 2 is a two-stage rocket, in use since 1974. It can put a 2- to 5-ton (1.8–4.5 MT) satellite into low Earth orbit. Long March 3 has an added recoverable cryogenic third stage using liquid hydrogen and liquid oxygen propellants. The most powerful rocket in this family, Long March 4, has a larger first stage than Long March 3 and a new third stage. It is capable of putting satellites into Sun-synchronous orbit. On one of its missions, in 1988, Long March 4 launched China's first meteorological satellite.

Japan's HI and HII Developed to launch large satellites, the three-stage, liquid-propellant HI launcher was 132 feet (40 m) high. It was capable of placing 7,100 pounds (3,200 kg) into low Earth orbit or 1,200 pounds (540 kg) into geosynchronous orbit. In 1986 HI launched an experimental *geodetic,* or Earth-measuring, satellite—one of 13 satellites it carried between 1986 and its final launch in 1992.

Designed to launch heavier, larger payloads than HI, the HII was developed for lunar and planetary spacecraft as well as multiple satellites. This powerful vehicle, which stands 164 feet (50 m) in height, is capable of putting a two-ton (1.8 MT) satellite into geosynchronous orbit. The HII is a two-stage launch vehicle, both

China's Long March (or CZ) family of launch vehicles has a high success rate and a per-satellite cost that attracts foreign customers. Many Long March rockets carrying satellites have been launched from the facility near Xichang.

The National Space Development Agency of Japan (NASDA) successfully launched the first HII rocket (above) from the Tanegashima Space Center on February 4, 1994.

stages of which use liquid oxygen/liquid hydrogen propellants. Two large solid-rocket boosters supplement the thrust of the first stage. The first burn of the second stage places the HII in low Earth orbit; a second burn places the payload into geosynchronous orbit. HII missions are planned throughout the 1990s.

India's SLV3 India's first successful rocket was the SLV3 (Satellite Launch Vehicle). SLV3 was a four-stage, solid-propellant rocket. Its first successful launch was in July 1980, when the satellite Rohini was placed in low Earth orbit. The SLV3 was also used successfully to put three satellites into low Earth orbit in the 1980s.

With the addition of two boosters, the SLV3 became the basis for the unsuccessful ASLV rocket, which was abandoned. India is now developing the PSLV (Polar SLV), which will be its first launch vehicle with two liquid-propellant stages; the PSLV also has two solid stages.

Israel's Shavit In September 1988 Israel launched Ofeq I ("horizon"), its first satellite, using the Shavit ("comet") rocket. This two-stage, solid-propellant rocket was derived from the medium-range military ballistic missile, Jericho II.

Ofeq I had an unusual launch. The rocket had to ascend in a westerly direction to avoid flying through the airspace of the Arab countries to Israel's east. This meant flying against the direction of Earth's rotation. A launch without rotational boost meant a smaller payload.

Israel is now interested in developing and launching a geosynchronous communications satellite. Because launching this satellite would exceed the Shavit's capabilities, a commercial launch will likely be arranged—possibly with Europe's Ariane rocket.

Japan's HII rocket, blasts off over the ocean near Tanegashima Space Center.

Launch Vehicle and Propulsion Concepts

Research and development of more efficient launch vehicles and better methods of propulsion are always in process. Scientists and engineers around the world continue to experiment and test, seeking more of the dramatic breakthroughs in knowledge and progress that have characterized the Space Age so far.

Shuttle Craft

Sänger/HORUS In 1985 West Germany planned to build a reusable spaceplane called Sänger, after Eugen Sänger, the engineer and physicist who conceived it. The spaceplane was envisioned as a two-stage system in which a carrier vehicle would lift a small upper stage—either a piloted shuttle-type spacecraft called HORUS (Hypersonic Orbital Reusable Upper Stage) or an unpiloted cargo carrier named CARGUS. Conventional rocket engines would put this upper stage into orbit, after which the carrier aircraft would return to its base. The HORUS shuttle would return to Earth, landing on a runway after completion of its mission. The CARGUS module would not be recovered.

Payload capacity to low orbit was planned to be either 9,000 pounds (4,080 kg) of cargo or 4,500 pounds (2,040 kg) of cargo and a crew of ten.

Sänger/HORUS was cancelled in the early 1990s due to budgetary constraints. Research is now focused on a crew transport vehicle to be used as a space station shuttle craft.

Hermes With strong support from France during the 1980s, ESA designed the reusable Hermes shuttle vehicle for transferring crews and equipment to space stations, servicing unpiloted platforms, and repairing satellites in orbit. Hermes was also intended to put scientific experiments into orbit during unpiloted flights.

An Ariane 5 rocket would have launched the Hermes. Plans called for Hermes to carry a three-person crew and 6,000 pounds (2,725 kg) of payload. Overall weight was to be about 42,000 pounds (19,070 kg). Hermes was cancelled, however, in the early 1990s for economic reasons.

Single Stage to Orbit Launcher: The Delta Clipper The experimental Delta Clipper takes its name from the Yankee clipper ships that opened sea trade routes more than 200 years ago.

Derived from the Thor-Delta launch vehicle, the Delta Clipper would have the capability of boosting 20,000 pounds (9,070 kg) to low Earth orbit using only one stage. Powered by eight rocket engines, the vehicle would be approximately 130 feet (40 m) in height and 40 feet (12 m) in diameter at its base.

The Delta Clipper takes off and lands vertically. Its on-board guidance system and rocket engines allow it to fly through winds and to land accurately on a small pad. After being towed back to the flight stand to unload cargo or passengers, the Delta Clipper would be serviced, refueled, and reloaded for the next flight. Although the exact timetable for development of this vehicle has not been set, to date there have been several test flights.

Heavy Lift Launch Vehicle An Advanced Launch System (ALS) was being developed that would have been capable of delivering payloads ranging from 1,000 to 220,000 pounds (450–99,800 kg) into low Earth orbit. The ALS project was cancelled, however, in the late 1980s.

In order to allow heavier payloads, parts of the launch vehicles were being constructed using a recently developed lightweight aluminum-lithium alloy. These heavy lift launch vehicles would have been used for NASA and Department of Defense missions, and were planned to have the capability of launching up to seven payloads on three flights within a five-day span.

During all phases of its flight (left to right)— launch, mid-flight, and landing—the experimental Delta Clipper maintains its vertical position. If successful, the Delta Clipper will be a breakthrough in low Earth orbit travel.

Propulsion Systems Research

Nuclear Propulsion Since the 1960s when the NERVA (Nuclear Engine for Rocket Vehicle Application) program proved the feasibility of nuclear propulsion, many nuclear propulsion concepts have been proposed. Included in these are nuclear thermal, nuclear electric, and nuclear pulse propulsion.

Nuclear thermal propulsion would involve the use of a nuclear reactor to heat a propellant fluid which expands and provides thrust. In nuclear electric propulsion, a fission or fusion reactor supplies electrical power to charge a propellant gas and accelerate it to a higher velocity than by simple heating. Nuclear pulse propulsion would propel the spacecraft using nuclear bomblets. Orion was an experimental model of this concept tested in 1968 and 1969, but it was abandoned for political and safety reasons.

Although nuclear propulsion could power spacecraft with marked savings in cost and travel time, safety concerns related to use of radioactive materials have not been adequately addressed to permit active research and development.

Solar Sailing A solar sail would propel a spacecraft with the pressure created by the stream of *photons* (tiny units of light energy) from the Sun. Once the spacecraft was in orbit, a lightweight aluminized plastic sail, miles long, would unfurl. Changing the position of the sail would increase or decrease speed. Although the thrust created by the photon stream is very low and interplanetary journeys would require years, solar sailing would be inexpensive because no fuel or engines would be required.

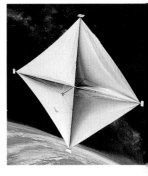

Scientists envision using the solar sail to enable spacecraft to fly in formation with celestial objects such as Halley's Comet. Launched from a Space Shuttle the sail spacecraft could study and photograph the speeding comet.

Laser Fusion Propulsion Laser propulsion could be at least 1,000 times as powerful as chemical propulsion and could make the entire solar system more accessible. Energy for propulsion of a spacecraft would be provided by a nuclear fusion reaction. To accomplish this, heavy hydrogen isotopes injected into the thrust chamber would be struck by a laser pulse as they reached the fusion point.

The Journey into Orbit

In October 1957, the "beep, beep" from orbit of Sputnik, the first satellite, was truly heard around the world. Once the United States had matched that Soviet achievement, the race between the two superpowers to launch humans into space was the next, daring step. Courageous pilots ventured into orbit in increasingly sophisticated spacecraft, testing their limits and endurance during ever longer missions.

Another stage in the
dream of space travel,
a U.S. Space Shuttle
moves gracefully over
the Earth's horizon.

First Satellites

Sputnik

As part of the 1957–1958 International Geophysical Year, both the Soviet Union and the United States planned to launch artificial satellites into Earth orbit. The Soviet Union stunned the world when it did launch the first satellite—Sputnik—on October 4, 1957, ushering in the Space Age.

Sputnik ("satellite") was a steel sphere, measuring 23 inches (58 cm) in diameter and weighing only 184 pounds (84 kg). Sputnik entered orbit and less than 100 minutes later was over its Baikonur launch site again, sending a "beep-beep" signal from its transmitter. The sound of Sputnik was followed all over the world by ham radio operators as well as scientists using very sophisticated equipment. Its low Earth orbit brought it as close as 142 miles (228 km) to Earth's surface and as far away as 589 miles (947 km). It completed a single orbit in 96 minutes and 17 seconds. Radio signals were transmitted to Earth through four metal antennas, 96 to 116 inches (2.5–3 m) long. For 21 days Sputnik transmitted data about the temperature inside the satellite to Earth. Its orbit decayed and on January 4, 1958, it burned up as it reentered Earth's atmosphere.

The Soviet rocket that launched Sputnik, and later Sputnik 2 and 3, was designed by Sergei Korolëv. With some changes, it was the predominant rocket in the Soviet Union for the first 25 years of the Space Age.

A Soviet engineer readies Sputnik for its historic journey. Once it was in orbit, scientists from Johns Hopkins University measured the frequency of Sputnik's radio signals from a single ground station.

A New Sound, A New Era

A U.S. radio announcer said of the "beep-beep" radio signal from Sputnik: "Until two days ago, that sound has never been heard on this Earth. Suddenly it has become as much a part of twentieth-century life as the whir of your vacuum cleaner . . . The satellite is still maintaining a speed of 18,000 miles an hour, a dozen times faster than any man has ever flown."

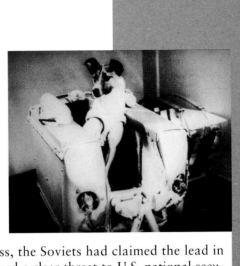

Laika rocketed into orbit aboard Sputnik 2, becoming the first animal in space. Her restraining belt kept her in place during liftoff and weightlessness. She waits in her seat to be placed in the capsule.

With Sputnik's success, the Soviets had claimed the lead in space technology and posed a clear threat to U.S. national security during the Cold War, when political tensions were running high. Within five days of the launch of Sputnik, a special committee of the Air Force Scientific Advisory Board was recommending that Intercontinental Ballistic Missiles (ICBMs) could serve as space boosters. The Air Force began to think of itself also as a space force. This led, in 1958, to the formation of the National Aeronautics and Space Administration (NASA).

Sputnik 2 Even as the reaction to Sputnik was rippling throughout the world, the Soviet Union launched a second satellite, Sputnik 2, on November 3, 1957. The first living creature—a female dog named Laika ("barker")—was placed in Earth orbit. The satellite was larger, about five times heavier, and more sophisticated than its predecessor. It had two compartments: one held instruments for recording temperature and pressure, and the other held a capsule for the spacecraft's passenger.

Monitoring equipment showed that Laika was affected minimally by the launch and weightlessness had no adverse effect on her. The satellite, however, was not capable of returning her to Earth, and a week later, when her air supply was gone, Laika died. Her body was incinerated when Sputnik 2 reentered the atmosphere in April 1958.

Carrying Sputnik, the Soviet rocket R-7 lifts off from Baikonur cosmodrome, or space center. The launch signified not only the beginning of the Space Race, but also the Soviet Union's initial domination of it.

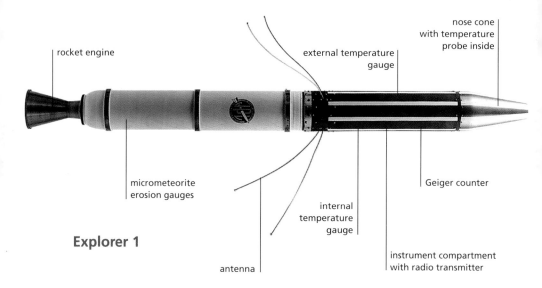

rocket engine

external temperature gauge

nose cone with temperature probe inside

micrometeorite erosion gauges

internal temperature gauge

Geiger counter

Explorer 1

antenna

instrument compartment with radio transmitter

Other Sputnik Highlights Sputnik 3, a metallic cone with solar cells along its sides, was launched on May 15, 1958. During its 691 days in orbit, it collected and returned data on micrometeorites and cosmic rays. This satellite weighed about 3,300 pounds (1,500 kg). Its launch sent a clear message that the Soviet Union had far more powerful rockets than the United States.

With the exception of Sputniks 7 and 8, Sputniks 4 through 10 were called Korabl-Sputnik ("ship-satellite") to convey the idea that, like a ship, they would return. These missions shared the goal of mastering the safe return of the satellite, and whatever living creatures were aboard, to Earth.

Sputnik 4 lifted off on May 15, 1960, on a Vostok launch vehicle. Because of a failed reentry attempt, Sputnik 4 remained in orbit for two years.

Sputnik 5 was launched on August 19, 1960, carrying two dogs, and mice, rats, houseflies, and plants. It completed 18 orbits of Earth and landed safely the next day. Sputnik 6 was sent up in December of the same year. It also carried a variety of animals, insects, and plants, but it burned up on reentry.

Sputnik 7 and Sputnik 8, planetary probe missions to Venus, were launched in February 1961, but were unsuccessful.

In March of 1961, Sputnik 9 was launched carrying various animals. After one orbit, it reentered the atmosphere and landed safely. Within a month, Sputnik 10 was launched. These two missions were dress rehearsals for the first piloted spaceflight around Earth that took place the following month.

Teams of NASA scientists, engineers, and technicians work on various components of Explorer 7 prior to its launch by Juno 2 on October 13, 1959.

Explorer Satellites

The first U.S. satellite, Explorer 1, was launched by the U.S. Army's Jupiter C rocket on January 31, 1958—almost 4 months after the launch of Sputnik. Explorer 1 was a bullet-shaped satellite, developed by a University of Iowa team led by Professor James Van Allen. It was only 80 inches (203 cm) long and weighed about 31 pounds (14 kg).

At its highest altitude, Explorer 1 reached 1,529 miles (2,460 km) above Earth, higher than either Sputnik or Sputnik 2. At this altitude, the satellite detected a zone of intense radiation inside Earth's magnetic field. This area later became known as the Van Allen radiation belts, which are composed of protons and electrons from the Sun, that surround Earth. Explorer 1 remained in orbit until 1967.

Some Explorer Scientific Missions Fifty-five satellites with the name Explorer were launched between 1958 and 1975. Between 1977 and 1984, ten more members of the Explorer family were launched, each named for the specific purpose of its scientific mission. These missions studied and measured a wide variety of space phenomena, including magnetic fields, solar wind, and ultraviolet radiation.

Explorer 6, launched in August 1959, gave the world the first television view of Earth. The satellite orbited as high as 16,373 miles (26,350 km) above the planet. The first spaceborne meteorological experiment was carried out by Explorer 7, launched in October 1959. The experiment was designed to detect and measure radiation in space.

Explorer 43's main mission was to study interplanetary space by conducting a radio astronomy experiment. Results of the mission confirmed those of Explorer 38: radio emissions were detected, probably emitted by electrons in space and spiraling under the influence of magnetic fields from the Sun and Jupiter.

Vanguard

In the early 1950s, the armed services sent proposals to the Defense Department for satellite programs. Von Braun's army team and a navy group sent one proposal, named Project Orbiter, which called for placing a 5- to 15-pound (2–7 kg) U.S. Navy satellite into orbit with the army's Jupiter C missile. A second proposal, made by another branch of the navy, was named the Vanguard Project. It called for a 20-pound (9 kg) satellite to be launched by a new three-stage rocket.

The Defense Department decided to proceed with Vanguard, in part because it could be shown to the world as a nonmilitary vehicle designed for peaceful purposes. President Eisenhower's policy was that Americans should demonstrate that rockets could contribute to humankind's welfare, as well as wage wars. Vanguard was designed to launch a scientific satellite during the International Geophysical Year (1957–1958).

A series of Vanguard launches were attempted between March 1958 and September 1959. The backup Vanguard 1 satellite was presented to the Smithsonian Institution in Washington D.C. The Vanguard 2 satellite is shown here.

In 1957 and 1958, two attempts failed; the Vanguard rockets exploded, in clear view of the world press, shortly after liftoff. They were called Vanguard TV3s (for test vehicle); number 1 was reserved for the first successful launch.

Vanguard 1 Launched on March 17, 1958, this became the second American satellite to orbit Earth. It was nicknamed "the Grapefruit" because it was a 6 inch (15 cm) sphere. It weighed only 3 pounds, compared to Sputnik's 184 pounds (84 kg). Reaching an altitude as high as 2,453 miles (3,947 km), Vanguard 1 carried temperature probes and two radio transmitters that allowed Earth stations to track its flight. The transmitters also enabled scientists to obtain data on Earth's shape and variations in its gravitational field. Using Vanguard data that continued to be transmitted for six years, geophysicists determined that Earth is somewhat pear-shaped.

Four failures followed before Vanguard 2 was launched in February 1959. Vanguard 3 was successfully launched in September 1959 after two more failed attempts.

Piloted Spacecraft

After Sputnik and Explorer 1, both the Soviet Union and the United States made parallel progress in the Space Race. The following summary of piloted spacecraft programs describes the successive developments in each nation separately.

Vostok

The Vostok ("east") spacecraft consisted of a cabin attached to an instrument module. In each Vostok mission, a single cosmonaut sat in the cabin in an ejection seat. A heat shield coated the spherical cabin to protect it from incineration during reentry. The instrument module, attached to the cabin by steel bands, contained a single, liquid-fuel retro-rocket and smaller attitude control thrusters.

Vostok 1 The first Vostok, weighing about 10,000 pounds (4,540 kg), was launched on April 12, 1961, with cosmonaut Yuri Gagarin aboard. Gagarin was the first human in space, a tremendous victory for the Soviets in the Space Race. The apogee of his orbit was about 203 miles (327 km) above sea level. After a single 108-minute orbit of Earth, Vostok 1's retro-rocket fired to begin reentry. The instrument section of the satellite was released and the capsule plummeted back toward Earth. At an altitude of about 1.5 to 2.5 miles (2.4–4.0 km), Gagarin ejected from the capsule and parachuted safely to the ground.

The first Vostok flight, in 1961, made cosmonaut Yuri Gagarin the first man in space. Vostok was designed to reenter Earth's atmosphere after ten days, even if the retro-rocket failed to fire, so Vostok missions always carried enough food for ten days.

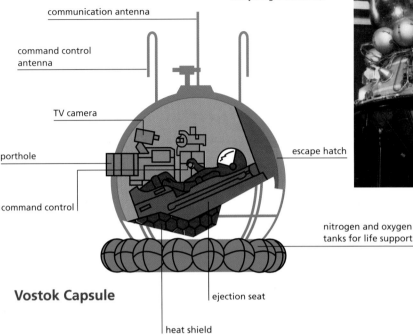

This close-up photograph shows the outside of the one-person Vostok capsule (detailed in the cutaway diagram below).

communication antenna

command control antenna

TV camera

porthole

command control

escape hatch

nitrogen and oxygen tanks for life support

Vostok Capsule

ejection seat

heat shield

Vostok 2 Launched in August 1961, Vostok 2 was piloted by Gherman Titov. His mission was to spend an entire day in space. This flight plan was required because as Vostok 2 orbited, Earth continued to revolve underneath it, and the spacecraft would no longer pass over the Soviet Union on every orbit. The mission had to be either less than five hours long or long enough to allow the cosmonaut to return to his own country. The decision was made that the flight would last for at least 24 hours.

On the third orbit, Titov ate some food paste. Within a few hours he took manual control and changed the attitude of the spacecraft. Titov then gained the distinction of being the first human to experience space sickness. He soon recovered, and his reentry was perfect. Titov ejected and landed safely by parachute 25 hours and 18 minutes after liftoff.

Vostok 3 and 4 Vostok 3, launched in August 1962, put Andrian Nikolaev into orbit. The next day, while he was still orbiting, Vostok 4 was launched with Pavel Popovich aboard. The two cosmonauts passed within 4 miles (6.4 km) of each other and spent the next few days talking with ground control and to each other. It was not possible to maneuver their spacecraft, so they slowly drifted apart. Simultaneous reentries were completed after Nikolaev was in space for four days and Popovich for three.

Vostok 1 cosmonaut Yuri Gagarin (left) and Vostok rocket designer Sergei Korolëv converse in the early 1960s. The son of a carpenter, cosmonaut Gagarin was awarded the highest honors of the Soviet Union—including the Order of Lenin.

Vostok 5 and 6 In June 1963, Valeri Bykovsky was launched in Vostok 5. Two days later, Vostok 6 was put into space carrying Valentina Tereshkova, the first woman in space. The two spacecraft came within about 3 miles (5 km) of each other, then they gradually drifted apart. By the time he returned to Earth, Bykovsky had made a five-day flight and set a new space endurance record. Tereshkova suffered from space sickness during her flight as other cosmonauts had, but she returned safely after a three-day flight. These were the last two Vostok flights.

Voskhod

The successes of the Vostok program took the Soviet Union into its next series of missions, the Voskhod ("sunrise") program. The Vostok spacecraft was redesigned to carry more than one cosmonaut. The Soviets successfully put a crew of three into space before the U.S. two-person Gemini mission was launched. Because Voskhod was a modification of the Vostok spacecraft, only one test mission, Cosmos 57, was required before launch.

Design Changes In modifying Vostok, the Soviets added a solid-propellant retro-rocket system to slow descent and allow cosmonauts to land inside the capsule, without ejecting. A reserve retro-rocket was added as a backup in case of a malfunction in the main system. The ejection seat was replaced with a couch for each cosmonaut turned in various directions to make them fit into the capsule. This meant that the crew would not be able to eject themselves if anything went wrong during launch or descent. The capsule was so crowded that space suits couldn't be worn, meaning that the crew would be unprotected in the event of a loss in air pressure.

Voskhod 1 On October 12, 1964, Voskhod 1 lifted chief pilot Vladimir Komarov, medical doctor Boris Yegorov, and design engineer Konstantin Feoktistov into space. This one-day mission made 17 orbits. Some biomedical data were reportedly collected, although little is known about this research.

Russian cosmonaut Aleksei Leonov trained rigorously for two years in preparation for his space walk. At 5 feet, 8 inches in height, the former MiG-15 pilot was too tall to be one of the first six cosmonauts. But in 1961, Sergei Korolёv chose him to make the first space walk.

Voskhod 2: The First Space Walk Voskhod 2 was launched on March 18, 1965; its objective was to achieve the first space walk. Pavel Belayev (commander) and space-walker Aleksei Leonov were aboard.

After entering an airlock, depressurizing it, and exiting through the outer hatch, Leonov drifted into space. He was connected by a safety line and a radio link to the spacecraft, and he wore a space suit with his air supply in a tank on his back. Floating in space for about 12 minutes, he photographed his spaceship and Earth and suffered no space sickness or vertigo. When he tried to climb back into the airlock, he found that his space suit had expanded during the space walk. By the time he reduced the pressure in his space suit enough to fit back into the airlock, Leonov had been in space for about 20 minutes.

This mission preceded the first two-person U.S. Gemini mission by five days, and this successful space walk, two months prior to the first U.S. space walk, reminded the world that the Soviets were still ahead in the Space Race. The next step was completing the new Soyuz spacecraft.

Voskhod 2 landed off course in a snowy forest where Belayev and Leonov lit a fire and spent the night with the wolves before they were rescued the next day.

Soyuz

Soyuz was first conceived as a vehicle for traveling to the Moon. In the early 1960s, Korolëv supervised the design of what was known as the "Soyuz complex," composed of three spacecraft: Soyuz A, B, and V. This project was canceled around 1964 in favor of a Moon flight using the more powerful Proton rocket, but the present Soyuz craft evolved from the Soyuz A.

The Basic Soyuz Spacecraft Soyuz means "union," an appropriate name for a spacecraft designed for docking. The original Soyuz craft, weighing about 15,000 pounds (6,800 kg), consisted of three modules: an instrument section with rocket engines and propellant, an orbital module, and a descent module.

After achieving Earth orbit, the cosmonauts would move into the larger orbital module and conduct experiments. The orbital module could be depressurized and serve as an airlock to allow the cosmonauts to walk in space. For the flight back to Earth, the crew would return to the descent module, which was equipped with a heat shield. If the descent module carried three cosmonauts, there was not enough room for the crew to wear protective space suits.

The three modules would separate upon reentry. Only the descent module would reenter the atmosphere intact; four solid-rocket motors would be fired to soften its landing. Forty Soyuz spacecraft (some unpiloted) were launched between April 1967 and May 1981, and the basic design of this spacecraft is still in use.

A technician works on the inside fittings of a Soyuz module on an assembly line. During the first few missions, crew members worked and rested in these orbital modules, which are located at the forward end of the spacecraft.

The Soyuz Missions

Soyuz 1–9 The first Soyuz, launched in April 1967, was designed to rendezvous with Soyuz 2. It was planned that two cosmonauts would transfer by space walk from Soyuz 2 to Soyuz 1 and return to Earth, but major attitude control problems developed on Soyuz 1. The launch of Soyuz 2 was cancelled immediately. In spite of desperate attempts to correct its problems, Soyuz 1 crashed when the landing parachute didn't open properly, and cosmonaut Vladimir Komarov died in the accident.

Soyuz 19 Spacecraft

orbital module |

descent module |

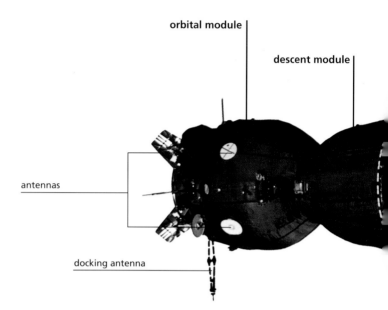

antennas

docking antenna

A Soyuz spacecraft and its launch vehicle are installed on the launchpad at Baikonur. Since the break-up of the Soviet Union, the launch site is now in the independent republic of Kazakhstan.

When Soyuz 2 was launched on October 25, 1968, it was unpiloted; Soyuz 3, with only Georgi Beregovoy aboard, was launched a day later. The two spacecraft approached within 650 feet (198 m) of each other but did not dock, then both returned to Earth. In January 1969, Soyuz 4 and 5 fulfilled the mission planned for Soyuz 1 and 2, docking and transferring crew members Aleksei Yeliseyev and Evgeni Khrunov by means of space walks. This was the first crew transfer in space.

Soyuz 6, 7, and 8 were launched in October 1969, one day apart. The seven cosmonauts on this triple mission performed maneuvers while Soyuz 6 carried out experiments. Soyuz 7 and 8 rendezvoused but did not dock. In June 1970 Soyuz 9's two crew members, Andrian Nikolaev and Vitali Sevastyanov, set a space endurance record of almost 18 days. It was reported that the cosmonauts found it difficult to readapt to Earth's gravity after that long period of weightlessness.

Soyuz 10 and 11 As the United States concentrated on the race to the Moon, the Soviets began to direct effort toward developing a space station. A modified Soyuz would be used to send crews to and from space stations. An added docking assembly would enable cosmonauts to move from a spacecraft to a space station without walking in space. On April 19, 1971, the world's first space station, Salyut 1, was launched. Soyuz 10, launched

solar panel

instrument and
equipment compartment

The descent module of a
Soyuz spacecraft, seen in
this simulator, contains
the main controls used by
the crew during launch,
descent, and landing. This
module was part of the
design for the joint 1975
Soviet–American flight.

four days later, achieved docking
with Salyut 1, but not crew entry. In
June Soyuz 11 docked, and the cosmonauts inhabited Salyut for three
weeks. On reentry, however, a tragedy occurred when a valve in the descent module opened. The air supply
rushed out of the cabin and into
space, and the cosmonauts—Georgi
Dobrovolsky, Vladislav Volkov, and
Viktor Patsayev—suffocated.

Postdisaster Adjustments As a result of the Soyuz 11 disaster, it was
decided that cosmonauts must wear pressurized space suits during critical phases of the flight, such as launch, docking, and
reentry. This meant that more room was needed in the capsule,
so one seat was removed. Soyuz missions were now limited to
two cosmonauts. The solar panels were also usually removed,
leaving the spacecraft dependent on batteries for electrical power.
If docking didn't occur on schedule, Soyuz craft would have to
return to Earth quickly, before the batteries ran down. Four of
the redesigned Soyuz craft retained solar panels. These spacecraft
made extended flights, including the Apollo-Soyuz international
mission in 1975 (Soyuz 19); Soyuz 13 and 22, which were scientific missions; and Soyuz 22, which observed Earth with a camera mounted on the orbital module.

Soyuz 12–40: Highlights The first flight of the redesigned Soyuz to accommodate two space-suited cosmonauts was Soyuz 12, launched in September 1973. The two cosmonauts tested the new life support systems, including the space suits. Soyuz 13 (1973) conducted astronomical research. Soyuz 15 (1974) carried out a night-landing test and stayed up only 48 hours. Soyuz 16 was a practice flight for the international Apollo-Soyuz Test Project, later flown by Soyuz 19. (See Apollo-Soyuz later in this chapter.) Soyuz 16 was launched in December 1974 with a new docking mechanism. Beginning with Soyuz 17, every flight except Soyuz 19 was a ferry to a space station. Most subsequent Soyuz missions were to the Salyut space stations.

Soyuz T The first of fifteen Soyuz T (Transport) missions was launched in December 1979. It was designed to dock with the Salyut station, and it included two solar panels to allow it to fly independently for four days if the docking wasn't successful. Soyuz T could accommodate three cosmonauts in space suits. It used a modified reentry procedure that required less fuel. The orbital module was discarded before the main engine was fired to return to Earth, reducing the total mass that would have to be decelerated for reentry. This made more fuel available earlier in the mission. The final Soyuz T mission was launched in 1986.

Soyuz TM Soyuz TM was a modified Soyuz T, with more advanced power supplies and parachutes and more space for equipment. First launched in 1986, the Soyuz TM will probably be used throughout the 1990s for transportation to and from Mir.

Soyuz Crews

As the Soviets developed their experience in longer-duration flights, they began to divide Soyuz missions into two categories: those with longer-duration crews and those with visiting crews. The former docked with the space station and remained there—often for periods of several months or as long as a year. Visiting crews would fly a fresh craft bringing food and other supplies to the station, then return to Earth in the old vehicle within about a week. The "new" spacecraft was left for the original crew.

The Mercury Missions

In 1959, NASA ordered its first piloted spacecraft, with the objective of placing astronauts in space, testing their reactions, and returning them safely to Earth. A total of 25 missions (test and flight) were launched in the Mercury program. The capsule was designed to be launched by either a Redstone or an Atlas rocket.

The original seven Mercury astronauts were an elite group of U.S. pilots: M. Scott Carpenter, L. Gordon Cooper Jr., John H. Glenn Jr., Virgil I. Grissom, Walter M. Schirra Jr., Alan B. Shepard Jr., and Donald K. Slayton. Each capsule, named by the astronaut that flew it, bore the number 7 in honor of the team.

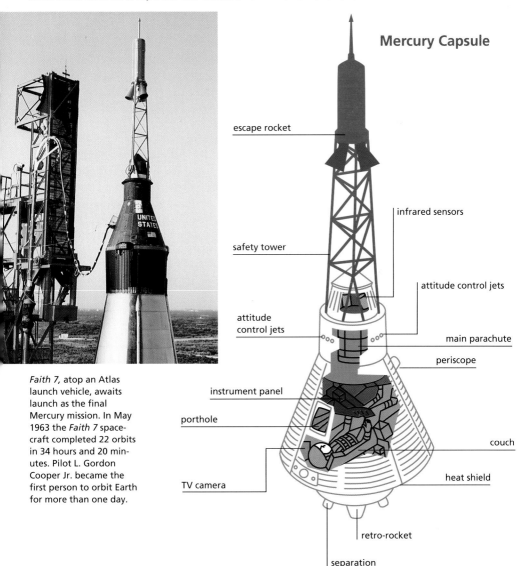

Mercury Capsule

escape rocket

infrared sensors

safety tower

attitude control jets

attitude control jets

main parachute

periscope

instrument panel

porthole

couch

TV camera

heat shield

retro-rocket

separation rocket

Faith 7, atop an Atlas launch vehicle, awaits launch as the final Mercury mission. In May 1963 the *Faith 7* spacecraft completed 22 orbits in 34 hours and 20 minutes. Pilot L. Gordon Cooper Jr. became the first person to orbit Earth for more than one day.

The Mercury Spacecraft The original design of the Mercury capsule was bell-shaped. It contained a custom-tailored couch for a single astronaut and a control panel. The cabin contained pure oxygen at about one-third the atmospheric pressure of Earth.

Attitude was controlled by 18 thrusters operated manually by the astronaut. Because the capsule had to be oriented in a particular way for reentry, the thrusters were designed to be operated in different combinations and by different control systems as a safety measure. Three retro-rockets would fire the capsule out of its orbit, but the capsule would not survive reentry if only one rocket operated properly. The outcome was uncertain if only two of them fired.

The primary challenge was to design a capsule that could survive the tremendous heat generated during reentry. The capsule was designed to enter with its blunt end first. This end was protected by a heat shield surrounding the base of the cabin, which would dissipate heat by slowly burning. Just prior to impact, the heat shield would detach from the base of the capsule and release a balloon that would inflate to cushion the astronaut's landing. A system for slowing the final descent of the capsule was also needed; a cylindrical area above the cabin housed the main and the reserve parachutes.

A safety tower supporting a solid-rocket escape motor was mounted on top of the capsule. In a launch emergency, the rocket would fire and lift the capsule from an explosion and parachute it into the ocean. Total height of the capsule, including its tower and the rockets attached to the heat shield, was about 26 feet (8 m). At launch it weighed about 17,500 pounds (7,900 kg).

Test Flights Mercury underwent seven suborbital test flights, five of which were successful; of its four orbital test flights, two were successes. In January 1961, a chimpanzee named Ham was put into a suborbital flight reaching an altitude of about 157 miles (253 km). During the flight, Ham performed simple tasks,

Freedom 7, ready to carry the first American into space, waits on the launchpad at Cape Canaveral on May 5, 1961. This was the third of the Mercury-Redstone missions.

Communications Blackout

As spacecraft reenter Earth's atmosphere from orbit, communications with Earth are cut off for a time because the friction of reentry heats the surrounding air and causes it to become ionized, or electrically charged. The ionized gas blankets the antennas, preventing the transmission or reception of radio signals.

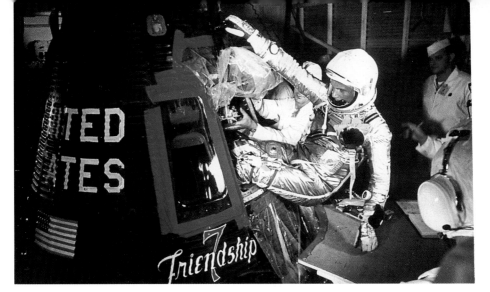

John Glenn climbs aboard *Friendship 7* at Cape Canaveral on February 20, 1962. Within hours he became the first U.S. astronaut to orbit Earth.

such as pulling a right-hand lever when a white light went on and a left-hand lever when a blue light was displayed. When the capsule splashed down, it began taking on water, but Ham was released unharmed. The Mercury capsule was deemed ready for its role as a piloted spacecraft.

Mercury's Suborbital Missions The first piloted Mercury flight *(Freedom 7)* lifted off on May 5, 1961. When Alan Shepard was launched to a speed of 5,146 miles per hour (8,282 km/hr) into suborbital flight, he became the first U.S. man in space. The trajectory apex reached was about 116 miles (187 km), and the horizontal "downrange" distance was about 303 miles (487 km). After the capsule separated from the booster rocket, Shepard took over the controls to test his ability to point the capsule in a specific direction. He descended safely in the *Freedom 7* capsule as it parachuted into the Atlantic Ocean after a flight lasting 15 minutes and 22 seconds. Although preceded by Yuri Gagarin's orbital flight less than a month earlier, this mission was an important and long-awaited step in the U.S. space program.

In July 1961 Virgil Grissom was launched on a suborbital Mercury flight in *Liberty Bell 7*. The flight was very similar to Shepard's mission until splashdown. While Grissom waited in the capsule for the recovery team, the bolts securing the hatch were triggered and water flooded in, forcing Grissom to abandon the capsule. As he floated in his space suit, Grissom realized that the suit was becoming waterlogged and was pulling him under. He struggled for several minutes to grab the sling that had been lowered from the rescue helicopter before finally reaching it and being pulled to safety. Attempts to retrieve the capsule by helicopter failed, and it sank to the bottom of the ocean, the only capsule not recovered.

The "Glenn Effect"

A little over an hour into his *Friendship 7* flight, John Glenn looked out the window at sunrise and saw brilliant, luminous specks in the sky. He first thought they were stars, then described them as looking like fireflies. The same year, on another mission, Scott Carpenter noticed the same phenomenon and tapped the interior wall of the capsule. When he did this, he saw hundreds of the particles outside the window. Experts concluded that the so-called Glenn Effect was frost emitted from the attitude control jets that was sparkling in the dawn sunlight.

An automatic sequence camera photographed John Glenn (above), weightless and traveling at 17,500 miles per hour (28,160 km/hr) aboard *Friendship 7*. The Earth (right) was also photographed from the spacecraft.

Mercury's Orbital Missions A Mercury capsule first orbited Earth on September 13, 1961, with an astronaut simulator mannequin. Launched by an Atlas rocket, this capsule landed safely after one orbit and was followed in November by a successful two-orbit flight carrying Enos, another chimpanzee.

Friendship 7 John Glenn was launched aboard *Friendship 7* on February 20, 1962—a day the United States had long waited for. After three minutes in orbit, Glenn began to feel himself becoming weightless. He told ground control, "Zero g, and I feel fine." After one hour and four minutes, he was scheduled to eat, and he emptied a tube of applesauce into his mouth. During the three-orbit flight, Glenn tested the attitude control system and described his observations of Earth from his perspective in space.

Though the mission went smoothly, it was not without drama. As Glenn prepared to begin reentry, mission control became aware that something was wrong. Initially Glenn was not informed of their concern, but when he received unfamiliar instructions, he knew there was a problem. Finally he was told that both the landing bag used to absorb the impact of a water landing and the heat shield might have come loose. The spacecraft was about to reenter the atmosphere at a speed that could turn the capsule into a fireball, and it was in danger of losing its heat shield. In addition, the shock wave from the friction-heated air would be at Glenn's back, creating temperatures five times that of the Sun's surface. Something had to be done.

Meanwhile, on the ground, a debate was in progress: should the retro-rocket be released, according to plan, or should it be kept in place in the hope that the straps of the rocket would also hold the heat shield in place? The decision was quickly made to leave it on longer than planned, and Glenn was told not to release the retro-rocket pack until he passed Texas. Glenn's reentry was particularly spectacular, from his vantage point, when pieces of the retro-rocket pack burned away and passed his window. He feared at one point that the spacecraft was burning and expected to feel the fire at any moment. The normal communications blackout as the spacecraft began its descent only compounded the apprehensions of those in mission control. But the heat shield remained in place, and Glenn splashed down safely, four hours and 55 minutes after liftoff.

Aurora, Sigma, and Faith In May 1962 Scott Carpenter made a three-orbit flight in the *Aurora 7* capsule. While essentially duplicating John Glenn's mission, he also carried out experiments and did a significant amount of maneuvering. *Aurora 7* landed about 250 miles (400 km) off target, and there was much tension at mission control until Carpenter and his capsule were located three hours after splashdown.

Sigma 7 was launched in October 1962, with Walter "Wally" Schirra as pilot. During Schirra's six-orbit mission, a telecast—the first ever—was beamed back to Earth.

In May 1963, Gordon Cooper was launched in the *Faith 7* spacecraft on a 22-orbit mission. The spacecraft orbited for 34 hours and 19 minutes. While in space, Cooper launched a sphere with strobe lights, which he was able to see on the following orbit. Thus, he became the first person to deploy a satellite from

Millions throng the streets of New York to stage a hero's welcome—as astronaut John Glenn rides down Broadway in a ticker tape parade.

orbital flight. When *Faith 7* developed a short circuit in its electrical system, Cooper was forced to make a manual reentry, but he landed within sight of the recovery ship.

America's Project Mercury ended well after logging a total of two days and six hours in space. From 1961 to 1963 six astronauts flew on six flights that demonstrated both safe spacecraft design and effective human performance in zero gravity. Project Gemini was ready to begin.

Gemini

In May 1961, when President John F. Kennedy made the commitment to put a U.S. astronaut on the Moon before the end of the decade, the United States had amassed only 15 minutes of piloted spaceflight. The Space Race became a race to the Moon, and the United States was determined to arrive there first. To help reach that goal, NASA increased its workforce from 16,000 in 1960 to over 33,000 in 1965.

The two-pilot Gemini project, following NASA's Mercury missions, was appropriately named for the constellation Gemini, "the Twins." Gemini includes the bright stars Castor and Pollux which are represented on the official design for the NASA patch (above).

Many technological challenges had to be met to accomplish a lunar mission. To control reentry and landing with more precision than was possible with the Mercury capsule, NASA decided to develop a two-pilot spacecraft. This spacecraft also had to be capable of spaceflights of up to two weeks. NASA designated the Gemini project to accomplish these goals.

Gemini was planned to perfect techniques needed for a lunar mission. Mission objectives were to put two persons in space, rendezvous and dock with another spacecraft, and have the astronauts walk in space. Meeting these objectives would determine whether Americans would indeed walk on the Moon within the decade.

The Gemini Spacecraft The Gemini spacecraft had the characteristic bell shape of Mercury. Including its reentry module, which housed the crew, and its wider adapter module, which housed the maneuvering rockets and oxygen and fuel tanks, Gemini was about 19 feet long (6 m). The spacecraft weighed 8,400 pounds (3,810 kg), over two and a half times the weight of the Mercury capsule.

The reentry module was a modified Mercury capsule, with only a 20-percent increase in the size of the capsule but with a 50-percent increase in cabin space. Two ejection seats sat side by side. A panel between them contained the controls for firing the small rockets to maneuver the spacecraft. A control panel was located in front of each seat. The Mercury hatch design was changed to allow opening and closing in space for the astronauts' space walks.

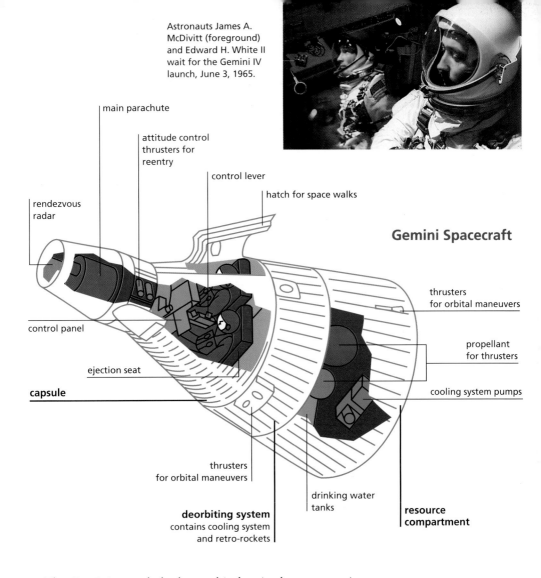

Astronauts James A. McDivitt (foreground) and Edward H. White II wait for the Gemini IV launch, June 3, 1965.

Gemini Spacecraft

main parachute

attitude control thrusters for reentry

control lever

hatch for space walks

rendezvous radar

control panel

ejection seat

capsule

thrusters for orbital maneuvers

propellant for thrusters

cooling system pumps

thrusters for orbital maneuvers

drinking water tanks

deorbiting system
contains cooling system and retro-rockets

resource compartment

The Gemini capsule had an orbital attitude maneuvering system (OAMS) with 16 thrusters that allowed the astronauts to maneuver the capsule in any direction and to accelerate or decelerate. Gemini was also equipped with a computer and a radar-tracking system. Electrical power was provided by either batteries for shorter missions or fuel cells for longer ones. The new fuel-cell system mixed hydrogen and oxygen and produced water as a by-product.

The adapter module contained four solid-propellant rocket engines that were fired in sequence to bring Gemini out of orbit and begin the trip back to Earth. Before reentry, the adapter module was discarded and incinerated as it fell and reentered Earth's atmosphere.

During his 22-minute space walk on the Gemini IV mission, astronaut Ed White wore an emergency oxygen supply chest pack. White and command pilot Jim McDivitt began the tradition of wearing American flag patches on their pressure suits.

At the back of the reentry module was the heat shield. Parachutes would lower the module into the ocean. The capsule would float upright after splashdown, allowing the astronauts to open the hatch while they waited for rescuers. Because Gemini was heavier than Mercury, the rescue plan called for U.S. Navy Frogmen to attach a flotation collar around the spacecraft before the hatch was opened. A helicopter lifted the astronauts from the floating capsule.

A Titan II rocket boosted the Gemini spacecraft into space. Safety features were added for carrying astronauts, including a malfunction detection system to inform the crew if any of the craft's vital systems were not functioning properly; extra electrical and hydraulic systems that could take over in the event of a main-system failure; and additional instrumentation to check the state of the rocket before launch and to monitor the spacecraft from the ground during flight.

The Gemini Missions

Gemini 1-3 The first two Gemini missions were unpiloted. In 1965 Gemini 1 remained in space for 64 orbits and confirmed the compatibility of the Titan II launch vehicle and the Gemini capsule. In 1966 Gemini 2, a suborbital flight, tested all systems, from liftoff to spacecraft recovery.

Gemini 3 was launched on March 23, 1965, with astronauts Virgil I. Grissom and John W. Young aboard. During the five-hour, three-orbit flight, the astronauts were able to alter their

**"It was the saddest moment of my life."
—Ed White describing his feelings
when he had to end his Gemini IV
space walk**

oval orbit, which ranged from 100 to 139 miles (161–224 km) in altitude, to a more circular one and to execute other orbital changes using the thrusters. These maneuvers were important practice for rendezvous to be carried out in later missions.

Gemini IV Launched on June 3, 1965, Gemini IV was a four-day flight, with astronauts James A. McDivitt and Edward H. White II. It generated enormous public interest because of the planned extravehicular activity (EVA) outside the spacecraft. The launch was broadcast to twelve nations in Europe, and millions watched. A rendezvous with the second stage of the Titan II launcher was their first objective, but the astronauts were unable to complete this difficult maneuver.

Gemini IV is probably best remembered for the first American space walk, carried out by White. Using a handheld maneuvering unit, he propelled himself for 21 minutes on his lifeline tether while traveling nearly 18,000 miles per hour (29,000 km/hr) in low Earth orbit.

Gemini V Gemini V was launched on August 21, 1965. Its mission lasted eight days and 128 orbits, briefly setting a world space endurance record. Astronauts L. Gordon Cooper Jr. (the first person to fly in space twice) and Charles Conrad Jr. piloted this flight. For the first time, the fuel cells needed for longer missions were used to supply electrical power, but they did not supply enough. This prevented some activities from being carried out, although the astronauts were able to rendezvous with an imaginary target.

Gemini VI and VII The launch of Gemini VI was scheduled for October 25, 1965, but the mission was cancelled when the Agena target vehicle with which the crew planned to rendezvous exploded during launch. This left Walter M. Schirra Jr. and Thomas P. Stafford without a target. NASA decided to launch Gemini VII first as a rendezvous target for Gemini VI.

Gemini VII was launched on December 4, 1965, and the mission lasted 14 days, the longest flight in the series. Its objectives were to study the physical effects of a long flight on astronauts Frank Borman and James A. Lovell Jr. and to act as a rendezvous target for Gemini VI. Gemini VI was launched 11 days later and immediately began to approach Gemini VII.

The Gemini IV spacecraft lifts off from the Kennedy Space Center—the first launch to be monitored around the world by means of the "Early Bird" satellite.

Making their planned rendezvous some 160 miles (257 km) above the surface of the Earth, Gemini VI crew members photographed the Gemini VII spacecraft through their hatch window.

The two spacecraft were able to move within 1 foot (30.5 cm) of each other—close enough for the astronauts to wave to one another through the windows. This first fully successful space rendezvous was accomplished in less than six hours after launch, and the two spacecraft continued to fly in formation for 20 hours and 22 minutes. The crews practiced rendezvous maneuvers and formation techniques that were vital to the success of future Apollo Moon missions.

On December 16, Gemini VI returned to Earth after 17 orbits, leaving Gemini VII in space for almost three more days. The Gemini VII astronauts spent 330 hours and 35 minutes in space, completing 220 orbits without long-term physiological problems. This mission proved that humans could endure lunar missions lasting up to two weeks.

Gemini VIII: Docking Achieved Launched March 16, 1966, Gemini VIII achieved the world's first docking, with an Agena target vehicle. The mission, however, unexpectedly lasted under 11 hours. About 20 minutes after the docking, astronauts Neil A. Armstrong and David R. Scott reported that the two spacecraft had begun spinning rapidly. Armstrong was able to temporarily stop the spinning using the thrusters, but it soon resumed, endangering both the spacecraft and Gemini VIII's astronauts. The astronauts felt that the problem was in the target vehicle, and mission control instructed them to separate from the Agena.

The Sandwich Subterfuge

A corned beef sandwich was smuggled onto Gemini 3 by one of the Gemini astronauts and was eaten by Virgil Grissom. The flight plan had called for John Young to eat a few specially prepared foods and for Grissom not to eat at all. After this infraction, stricter rules were passed about what astronauts could take onboard.

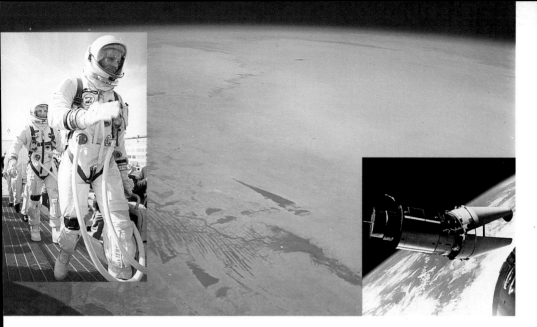

After undocking, however, Gemini VIII began to spin even faster—about one revolution per second. Armstrong and Scott soon began to have trouble seeing clearly, and the violent motion of the capsule made communications with the ground nearly impossible. The astronauts were approaching unconsciousness, which, if the capsule continued its motion, could be fatal. As a final attempt to save their lives, Armstrong broke a mission rule: he turned off the thruster system and activated the reentry control system. The hand controllers began to respond and Armstrong and Scott gained control of their spacecraft again, but too much fuel had been expended to complete the planned three-day mission. They were forced to reenter and make an emergency landing in the Pacific Ocean. Later it was discovered that a maneuvering thruster on Gemini VIII had become stuck in the open position, causing the spacecraft to spin.

Gemini IX Gemini IX was launched on June 3, 1966. Astronauts Thomas P. Stafford and Eugene A. Cernan had been assigned to the mission after the originally slated astronauts Elliot M. See and Charles A. Bassett died in a plane crash. Gemini IX, planned as a three-day, 45-orbit mission, was unable to complete its intended docking because the protective cover on the Agena docking target remained partly attached.

During this mission, NASA became aware of potential problems that could arise during EVA. In struggling to put on his backpack maneuvering unit during his space walk, Cernan breathed heavily and his faceplate became fogged. The moisture couldn't be handled by the space suit air conditioner, and limited vision forced him to terminate the space walk about an hour later without testing the jet-powered backpack.

The Gemini IX astronauts (left), arriving for pre-launch exercises, look as if they are walking toward the beautiful images of African terrain (center) photographed by Gemini VI. Plans for a Gemini IX space docking failed when the rendezvous target, dubbed the "angry alligator," (inset) failed to open fully.

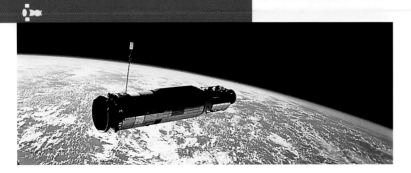

Gemini X Gemini X, with astronauts John W. Young and Michael Collins, was launched on July 18, 1966. A successful rendezvous was achieved with an Agena target vehicle five hours and 21 minutes after launch. After docking, the Agena's propulsion system sent both spacecraft to a new record altitude of 458 miles (737 km) high.

During the 15th orbit, with the hatches open, Collins photographed the stars in ultraviolet light, something impossible to do below Earth's atmosphere. He later began taking color photographs, but an irritant in the oxygen system temporarily blinded the astronauts, making it necessary for them to close the hatches and repressurize the cabin.

On the last day of the three-day mission, Gemini X undocked from the Agena and rendezvoused with another Agena left in orbit by Gemini VIII. In a 39-minute EVA, Collins propelled himself to the Agena with the handheld maneuvering unit. There he collected two scientific packages and returned to his spacecraft. Splashdown within sight of the recovery ship ended this complex, tremendously successful mission.

Gemini XI On September 12, 1966, Gemini XI was launched. The capsule rendezvoused and docked with the Agena target vehicle within the first orbit. This mission broke the altitude record set by Gemini X when the Agena propelled the craft to an altitude of 853 miles (1,373 km). From that vantage point, astronauts Charles Conrad Jr. and Richard F. Gordon Jr. viewed a breathtaking sight—more than 17 million square miles of Earth beneath them, including Australia and Southeast Asia.

The second day of the flight, Gordon initiated an EVA to find out how the Agena would react after undocking. As a result of the physical exertion of this EVA, his faceplate fogged up and, like Cernan on Gemini IX, he had to terminate the space walk and return to the capsule. Later the astronauts "created" gravity (1.5 thousandths of normal Earth gravity) by tethering the two spacecraft and letting them slowly rotate around each other. This was the first time artificial gravity had been created in space.

Their landing was the first automatic splashdown of the Gemini program. After the retro-rockets fired, the computer made the adjustments until the spacecraft landed in the ocean.

"The End"—Words worn across the backs of Lovell's and Aldrin's space suits as they boarded the final Gemini spacecraft, Gemini XII.

Gemini XII The final mission of this series, Gemini XII, lifted off on November 11, 1966. About four hours after launch, astronauts James A. Lovell Jr. and Edwin E. Aldrin Jr. rendezvoused with their Agena target vehicle.

The most important task in this mission was an extended EVA carried out by Aldrin. Because of the problems encountered during the flights of Gemini IX and XI, NASA decided that Aldrin should take a less strenuous space walk. Putting his underwater training skills to work, Aldrin completed about 20 simple tasks, many related to future repair of spacecraft. After a space walk lasting two hours and twenty minutes, and two stand-up, hatch-open periods, Aldrin had set a new world record for EVA, about five and a half hours. Four days and 63 orbits after launch, the last Gemini mission made its reentry and splashdown.

The ten piloted Gemini missions of 1965 and 1966 included the first U.S. space walks and a series of successful rendezvous and docking exercises. The 16 Gemini astronauts (four of whom flew twice) proved that longer-duration spaceflight could be accomplished safely. In spite of new hardware and difficult maneuvers, Gemini continued the perfect safety record of the Mercury missions. From the first Mercury flight to the final Gemini mission, totalling about 18 million miles (29 million km), no astronaut lost his life. The Apollo program would build on what had been learned and head for the Moon. (See Chapter 5.)

Gemini XII (top right) plummets toward splashdown in the Atlantic Ocean, some 600 miles east of Cape Canaveral. Astronauts James Lovell and Edwin Aldrin celebrate the success of the final Gemini mission aboard the recovery ship *USS Wasp* (right). This mission brought the total Gemini time in space to about 2,000 hours and completed a perfect safety record.

Space Stations

The usefulness of a permanently occupied space station was considered before the Space Age, for example, as a way station for missions to the Moon and planets. Both the United States and the Soviet Union also recognized the value of a permanent presence in space for purposes including Earth observation for military and scientific programs, and for research. Once the United States had won the Moon race, the major emphasis of the Soviet space program shifted toward the development of Earth-orbiting space stations.

The Salyut Space Stations

The Soviet Union's Salyut ("salute") program included space stations designed for both civilian and military purposes. Salyut 1 was launched April 19, 1971, nearly two years before the United States launched its first space station, Skylab. Salyut weighed over 25 tons (23 MT), was 47 feet (14.3 m) long, and had a diameter of 13 feet (4 m) at its widest point. It was a series of four cylinders of varying diameters and lengths. Power was supplied to the station by four solar panels.

Eventually, piloted Soyuz spacecraft and unpiloted Progress and Cosmos spacecraft hooked up to Salyut, adding more living space. Visiting crews entered the space station through a transfer compartment at one end. The connected work compartment also accommodated the crew during off-duty periods and was the location of the main control center. The next cylinder housed a treadmill, sanitation facilities, and most of the experiments. Finally, there was a cylinder containing the propulsion system for modifying Salyut's orbit and providing attitude control.

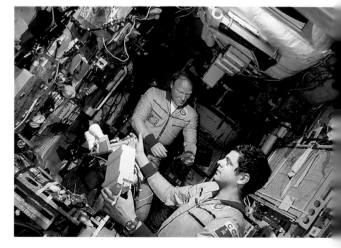

During their record-setting 237 days aboard Salyut 7 space station, cosmonauts from Soyuz T-10 carried out notable experiments in space medicine and other fields, even though the station was plagued by electrical problems requiring frequent repair.

Salyut 1 The first cosmonauts to dock with Salyut were launched in Soyuz 10 on April 23, 1971. For unknown reasons, they docked but never entered the station. Instead, they undocked and returned to Earth on April 24. On June 6, a second crew was launched in Soyuz 11, docked successfully, and Georgi Dobrovolsky, Viktor Patsayev, and Vladislav Volkov spent over three weeks in Salyut 1. But, tragically, on the return to Earth, a valve accidentally opened, releasing their air into space, and the crew quickly blacked out and suffocated. After six months Salyut 1 burned up as it reentered Earth's atmosphere.

Salyut 2 Salyut 2 never stabilized in orbit. It reentered Earth's atmosphere in April 1973, the same month it was launched.

Salyut 3 Salyut 3, a military station, was placed in a low orbit during the last week of June 1974. On July 3, Pavel Popovich and Yuri Artyukhin docked Soyuz 14 with Salyut and stayed aboard the space station for two weeks. In August, Gennadi Sarafanov and Lev Demin were launched in Soyuz 15. Their rendezvous was unsuccessful. In 1975 the empty station burned up in the atmosphere.

During its four years in operation, Salyut 7 was host to crew members from other countries—including a French cosmonaut in 1982 and one from India in 1984. Salyut 6 also had visitors, from Asia, Eastern Europe, and Cuba.

Salyut 4 A civilian mission, Salyut 4 was launched on December 26, 1974. This station was similar to Salyut 1, although it was equipped with a solar telescope. On January 10, 1975, Aleksei Gubarev and Georgi Grechko visited Salyut 4 from Soyuz 17. They stayed on the station until February 9, completing what was then the longest (one month) Soviet spaceflight.

Another crew was launched in April 1975 for Salyut 4, but they were forced to cut their mission short. The following month, Pyotr Klimuk and Vitali Sevastyanov were launched in Soyuz 18. Their mission aboard Salyut 4 lasted 63 days.

A final unpiloted mission was made to Salyut 4 when Soyuz 20 was launched on November 17, 1975. The primary purpose of this last mission is believed to have been the testing of new docking systems. Salyut 4 remained in space until February 1977.

Salyut 5 Launched on June 22, 1976, Salyut 5 was the final military station of this series. Its initial crew, Boris Volynov and Vitali Zholobov, were launched on Soyuz 21 on July 6. It is believed that they returned to Earth prematurely because of air supply problems. The next crew, Viktor Gorbatko and Yuri Glazkov, was launched February 7, 1977, on Soyuz 24; they corrected the air supply problem and stayed aboard until February 25. The station reentered Earth's atmosphere on August 8, 1977, and burned up.

This close-up photograph of the Skylab space station—shown over the Amazon River valley of Brazil—was taken from the command module during maneuvers prior to docking. Skylab 3, which carried the space spiders, Anita and Arabella, included many other scientific and medical experiments in space, as well as tests involving mice and minnows.

Salyut 6 and 7 Salyut 6 and 7 were "second-generation" Soviet space stations with the same basic design as Salyut 4, but with a second docking port. The change made it possible for automatic resupply missions to restock Salyut. Salyut 6 and Salyut 7 missions were noted for long endurance records, complex scientific research, space station repair, and visits by guest cosmonauts from other countries.

Salyut 6 was launched on September 29, 1977. In addition to the added docking port, it had a new propulsion system, which could be refueled while in orbit. Salyut stayed in space for about four years. During that time, 31 spacecraft docked with the space station and unpiloted Progress spacecraft transported supplies to it before its reentry in 1982.

Salyut 7, launched April 19, 1982, contained many additional modifications. Among the crew members who visited was Svetlana Savitskaya, in 1984, the first woman to complete a space walk. Also in 1984 the crew of Soyuz T-10 set a new space endurance record aboard Salyut 7—237 days in space. Last occupied in March 1986, Salyut 7 fell back to Earth in 1991.

The Skylab Space Station

Launched on May 14, 1973, two years after the launch of Salyut 1, Skylab has been the only U.S. space station to date. Three crews inhabited Skylab for a total of 171 days and 13 hours and performed hundreds of experiments.

The station was 118 feet (36 m) long, had a diameter of 21 feet (6.4 m), and weighed almost 100 tons (91 MT). Skylab consisted of a two-level workshop (about the size of a small house) including living quarters and work space, a multiple docking adapter/airlock module, and the Apollo Telescope Mount, a solar observatory.

The official design for the Skylab patch (below) depicts the U.S. space Station in Earth orbit with the Sun in the background. As part of its mission, Skylab's solar telescopes increased scientific knowledge of the Sun and the multitude of solar influences on Earth's environment.

Astronaut Owen Garriott (above), science pilot for Skylab 3, works at the telescope console in Skylab's experiment control center.

> In space, an astronaut's heart may beat as infrequently as 30 times per minute during sleep; at rest on Earth, the adult heart averages 70 beats per minute.

Life on Skylab Living conditions in Skylab were better than those on Apollo spacecraft, with more spacious living area and private sleeping quarters. The astronauts were able to sleep more (6 to 8 hours per night) in sleep restraint bags, hung on the wall.

Dining conditions were also better than those of any other spacecraft. Because weight and space restrictions are not as tight on the space station, a freezer held a variety of 72 different food items. Astronauts ate at a small table near a window where they could enjoy the view.

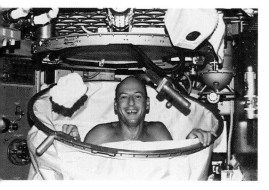

As Skylab orbits the Earth, astronaut Charles Conrad, mission commander of Skylab 2, relaxes after a warm shower in the crew quarters of the orbital workshop. Conrad celebrated his 43rd birthday—June 2, 1973—while in orbit.

Preparing a meal on Skylab was a fairly simple matter. The astronaut with galley duty assembled the day's foods and added water to those that were dehydrated. The containers were then placed in a tray with heating elements, to be warmed for consumption at mealtime.

Fitness programs were developed as NASA learned that space exercise could help to counteract some of the effects of extended spaceflights, including loss of cardiovascular conditioning and muscle tone. Exercise equipment on Skylab included a stationary exercise bicycle and a modified treadmill. The bicycle could be pedaled with either the feet or the

Floating Sweat?

When astronauts exercise in weightlessness, the sweat doesn't drip off their bodies. Rather, it becomes a puddle that "just sort of slithered around," according to Skylab 4 astronaut William Pogue. Immediately after exercise, astronauts must catch the flying puddle and towel it up before it flies off and causes problems in the spacecraft. At right, Paul Weitz, pilot for the first crew, tries out some of the Skylab fitness equipment.

hands, but even with their feet secured in the pedals, astronauts found that a ceiling pad was necessary because they would sometimes float off the seat. Crews exercised while wired with electrodes and blood pressure cuffs to monitor their vital signs.

In addition, Skylab was equipped with a machine called the lower-body negative-pressure experiment. In weightlessness, fluid accumulates in the upper body. This machine compensates for this by pulling fluid from the upper body down to the legs.

Skylab also contained the first space shower, which was effective once the space-related quirks of the process were mastered. The space station also had the first private space toilet. It was equipped with seatbelts so the astronauts wouldn't float off.

Missions to Skylab Almost immediately after Skylab 1's launch by a Saturn V rocket (the last Saturn V launch), problems developed. A meteoroid shield was accidentally deployed during launch and was torn away. It also tore off one of the two main solar panels, and a piece of the shield jammed the second solar panel. Skylab 1's thermal shield was gone, and its main power generation system wasn't fully functional. With only the observatory's solar panels still functioning, the temperature inside the space station rose to over 125° F (52° C). Equipment damage and noxious gases became real threats. After an intensive ten days of problem solving, the first Skylab crew was launched to try to repair the damage.

Crew #1 When Skylab 2 mission crew members Charles Conrad, Paul Weitz, and Joseph Kerwin were launched on May 25, 1973, their first task was to try to unjam the solar panel, but the attempt was unsuccessful. Testing the cabin atmosphere for fumes and finding it safe, they entered the station. Though they

Near the end of the Skylab 2 mission, Charles Conrad and Paul Weitz went outside in pressure suits. Conrad retrieved solar telescope film canisters. Data gathered on the mission was stored aboard the command module and returned to Earth.

Hazards of Space Debris

Thousands of pieces of this debris—many of them microscopic—are already in orbit. Traveling at speeds up to 17,500 miles per hour (28,160 km/hr), they can seriously damage spacecraft and threaten the safety of the astronauts.

How large must a piece of space debris be to be dangerous? During the seventh Space Shuttle mission, the crew discovered a pit about 0.2 inches (0.5 cm) wide on one of the windows. Laboratory tests on Earth later suggested that the particle that caused the damage was a speck of paint from a satellite. If traveling at high speed, a similar particle of an inch or two in diameter could damage a spacecraft and endanger its crew.

Broken pieces of rocket boosters, pieces of old satellites, even lost clothing and tools—this human-made space trash is accumulating rapidly. As these remnants of spaceflight missions increase in number, the probability of their colliding with each other also increases. When these collisions occur, they create even more pieces of fractured trash.

Objects in low-Earth orbit eventually fall, but, depending on their altitude, may stay up for months or years. Space trash in geosynchronous orbit—that is, 22,300 miles (35,880 km) above the Earth's equator—will remain there for centuries.

were able to lower the temperature in the station by pushing a sunshade through an airlock, they still had to contend with a shortage of electrical power. During a three-and-a-half-hour space walk two weeks later, the men cut the strap holding the solar panel shut and restored power. This allowed them to carry out the planned experimental program during the remainder of the mission and to set a new space endurance record of 28 days.

Crew # 2 On July 28, 1973, the Skylab 3 mission put Alan Bean, Owen Garriott, and Jack Lousma in the space station for 59 days. Their work included further repairs, Earth and solar observations, and tests of a space-walk maneuvering unit. The crew also conducted a fascinating "classroom in space"—including a demonstration of zero gravity.

Crew # 3 The third and final crew to inhabit Skylab set the U.S. space endurance record of 84 days in orbit. During this successful Skylab 4 mission, launched on November 16, 1973, crew members Gerald Carr, William Pogue, and Edward Gibson photographed and studied the comet Kohoutek and a giant solar flare. Medical and material science experiments investigated the

Space Spiders

Does weightlessness affect how a spider spins its web? To answer this question, posed by a Massachusetts student, Skylab carried two special passengers—spiders Arabella and Anita. Arabella, allowed to begin to spin a web almost as soon as she arrived in space, was unsuccessful for the first two days, after which she was able to spin fairly normal webs. Anita, not allowed to begin for a few days, spun normal webs almost immediately.

Astronaut Edward Gibson floats above his crewmate Gerald Carr in the forward experiment area of Skylab, demonstrating the effects of zero gravity. The third crew member, pilot William Pogue, snapped this on-board photograph during the record-setting Skylab 4 mission.

unique conditions of weightlessness. Two additional records were set on this flight: the longest EVA—seven hours and one minute—and the greatest total EVA time for a single mission—22 hours and 21 minutes.

On this mission, scientists had an opportunity to study the physiological effects of a long flight on the crew. They also learned that astronauts have limits to their good temper; the crew briefly went on strike against what they perceived as excessively demanding schedules and requests from mission control. This confirmed that the psychological, as well as physical, well-being of space travelers is an important factor, particularly on longer missions.

By the end of this third and final visit to Skylab in February 1974, Skylab had made about 3,900 orbits of Earth, each taking 93 minutes. In 1979, Skylab fell back to Earth, and parts of it landed in the Indian Ocean and on the Australian continent. For the greatest percentage of its life, the space station was not inhabited; astronauts occupied it for only six of its 75 months in orbit. Nevertheless, Skylab missions were extremely productive.

Apollo astronauts caught a glimpse of their Soviet neighbors-in-space during the Apollo-Soyuz mission in July 1975.

Apollo-Soyuz Test Project

Early in the 1970s the improving political situation between the United States and the Soviet Union made a joint space mission possible. This mutually beneficial mission would keep NASA's astronaut team together between the end of the Skylab program and the beginning of the Space Shuttle program. It would also be a concrete representation of the détente declared by President Richard Nixon and Soviet Premier Leonid Brezhnev.

In October 1970, joint working groups began to meet to discuss technical and safety issues that would arise from a docking between an Apollo and a Soyuz spacecraft. At the superpower summit in May 1972, both sides signed an agreement to participate in a joint mission that would be launched in July 1975.

Prelaunch Preparations Because the spacecraft docking systems of the Soviet Union and the United States were not compatible, the most immediate need was to develop a new docking module with an Apollo docking feature at one end and a Soviet-compatible feature at the other. This module was carried in the Saturn 1B launch vehicle along with the Apollo spacecraft for use in orbit. It would also act as an airlock—astronauts or cosmonauts switching vehicles would spend time in this transfer compartment to become gradually accustomed to the atmosphere of the other vehicle. The U.S. spacecraft used a pure oxy-

Cosmonaut Aleksei Leonov and astronaut Donald Slayton greet each other in the Soyuz orbital module after the docking of Apollo-Soyuz. Soyuz crew members photographed the nearby Apollo (below) and its newly developed docking module.

gen atmosphere, while the Soviets used a higher-pressure oxygen and nitrogen mixture. If the Soviet crews tried to enter the American spacecraft without time in the airlock, they would likely suffer from "the bends," a painful, dangerous condition known to deep-sea divers who surface too quickly.

Other issues to be resolved included a mutual radio link system and the lack of a common language to be used for the mission. Each crew agreed to speak to their control centers in their own language and to try to communicate in each other's languages when possible. The Soviets would refer to the mission as Soyuz-Apollo and the Americans would call it Apollo-Soyuz. In the years preceding the launch, crew members and international technical experts met several times in both countries.

Soyuz 19 and Apollo America's crew included commander Thomas Stafford (of Gemini VI, IX, and X) with Vance Brand and Donald "Deke" Slayton, one of the original Mercury astronauts. Aleksei Leonov, the first man to walk in space, was the Soviet commander. The other Soviet crew member was Valeri Kubasov, of the Soyuz 6 mission.

Both launches went smoothly. Soyuz 19 lifted off on July 15, 1975, and Apollo followed about seven hours later. Apollo had the responsibility of tracking and rendezvousing with Soyuz 19. By the second day, the two spacecraft were in the same orbital plane, ready to rendezvous. That evening, July 17, 1975, when Soyuz 19 was in its 35th orbit and Apollo was in its 29th, the two spacecraft docked successfully. Stafford and Slayton entered

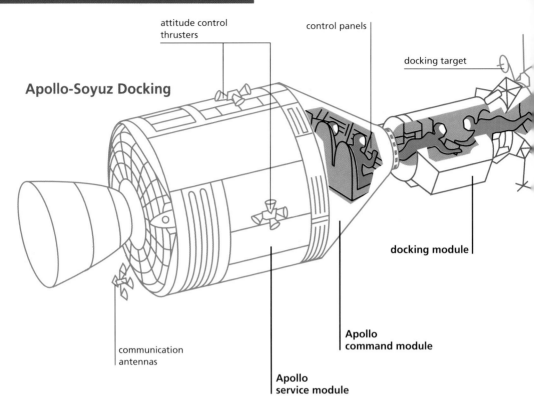

Apollo-Soyuz Docking

attitude control thrusters

control panels

docking target

docking module

Apollo command module

Apollo service module

communication antennas

the intermediate docking module and began adjusting to the pressure change of the Soviet spacecraft. Then, on live television, the two crews met and shook hands. It had taken almost five years to reach this point. The astronauts and cosmonauts exchanged greetings; messages were relayed to the crews from Soviet leader Leonid Brezhnev and President Gerald Ford.

The two spacecraft remained docked for two days. The crews visited each other's craft four times, and some joint experiments were carried out in astronomy, Earth observation, and metallurgy. They undocked and used each other's spacecraft to perform additional maneuvering exercises, docked again, and then separated for the last time.

Soyuz 19 returned to Earth on July 21, and Apollo spent three more days (nine total) in space. After splashdown, an astronaut error caused the fumes of fuel being dumped overboard to be drawn into the capsule. The fumes caused Brand to lose consciousness, but no permanent harm resulted. The crew probably saved their own lives by using their oxygen masks.

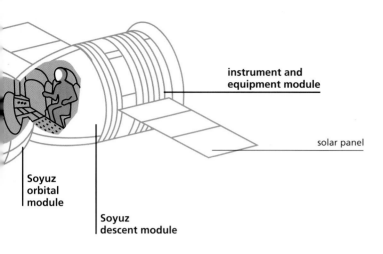

instrument and
equipment module

solar panel

Soyuz
orbital
module

Soyuz
descent module

The Mir Space Station

Mir ("peace," or "world") is currently the only operational space station. It is a descendant of the highly successful Soviet Salyut series of the 1970s and 1980s. Mir, the central component of a modular space station, was launched on February 20, 1986, by a Proton rocket. With its multiple-position docking system, it is a very versatile space station, capable of docking with piloted spacecraft, unpiloted Progress cargo vehicles, and various scientific modules. Its orbital altitude ranges from about 186 to 248 miles (300–400 km).

Though Mir is about the same size as the Salyut 7 space station, which it replaced, it does not carry as much scientific equipment. It is viewed primarily as the future living quarters of a large space station complex. The Salyut stations combined living and working quarters, but Mir's design responded to cosmonauts' demands that those areas be separated. Mir's environment provides more privacy and comfort than any spacecraft before it. Up to six cosmonauts can be accommodated in the space station, but normally only two or three occupy it except during crew changes.

Mir consists of four individual compartments, from front to back: the docking compartment, the work compartment, the living space, and the propulsion compartment. To these are added specialized modules for research.

Engineers at the assembly and testing shop of the Baikonur cosmodrome prepare Mir space station for launch. Mir has remained in operation since 1986.

The first cosmonauts to work on Mir after its launch in 1986 were Leonid Kizim and Vladimir Solovyev, who had also been aboard Salyut 7 during its record-breaking mission. After 50 days, they transferred back to the older space station—the first time crews had moved from one space station to another.

The Docking Compartment Mir's docking compartment is equipped with five docking ports. One port receives the Soyuz spacecraft, which transfers new crews to the station; the other ports are used for attaching modules for scientific and technological experiments. In addition to receiving incoming vehicles, the docking compartment contains television equipment and the station's electric-power-supply system.

Progress vehicles supply water and food to Mir and collect materials from the experiments carried out on board. These include astronomy observations, materials processing, Earth observations, and military research.

The Work Compartment The work compartment houses the main control console for operating the station, with systems for attitude control, navigation, and communications. On the opposite side of this compartment are the solar panels that, along with batteries, provide electricity.

A Ticket to Mir

Several paying civilian passengers have taken trips to Mir. The first was Toyohiro Akiyama, a Japanese journalist who traveled to Mir aboard Soyuz TM-11 in December 1990 and spent a week on the space station.

The Living Space The crew's living space—a common area and two small cabins with sleeping bags attached vertically to the walls—is connected to the work area. Exercise equipment, a work table, a dining area, the water supply, a sink, and a toilet make up the rest of the living area. Mir also has a water recycling system. At this end of the station is an airlock through which cosmonauts enter when docking occurs.

This photograph of Mir as it orbits above Earth clearly shows the solar panels that help provide electricity. Each panel has a surface area of about 250 square feet (76 sq m). Automatic controls adjust the panels to catch the greatest solar radiation.

The Propulsion Compartment The propulsion compartment, unlike the other three compartments, is not pressurized. It houses two rocket motors used for orbital adjustments and 32 attitude control jets. It also contains fuel reservoirs, the heating system, and the sixth docking port. Unpiloted Progress vehicles dock and transfer fuel from this port. A separate system handles regulation of Mir's interior atmosphere. Mounted outside the propulsion compartment is a directional antenna that is part of the satellite communications system with Earth.

The space station's position can be maintained either automatically or manually with computerized optical sensors, gyroscopes, and visual orientation instruments. Mir's size and low orbit make it visible from Earth and subject to the pull of Earth's gravity. Periodically rocket engines must be fired to propel the space station back to a higher altitude. Two primary motors toward the back of the propulsion compartment are used to make these orbital adjustments.

Mir Modules

Since Mir was launched, three experiment modules, customized for scientific and technological research, have been added to the Mir core. Each has its own heating system, engineered for the type of experiments being performed.

The first module, Kvant ("quantum"), launched on March 31, 1987, is an astrophysical observatory including ultraviolet, X-ray, and gamma ray telescopes. It also contains equipment for producing biological substances. Kvant 2, known as the "reequipment module," was launched on November 26, 1989. It contains a research section and the primary airlock for space walking. The Kristall module, which arrived at Mir on May 31, 1990, has two sections: an instrument compartment containing equipment for materials and medical research, and the docking compartment with two newly designed hatches that can accommodate *Buran*, if it becomes an operational space shuttle. Two additional modules, Spectr ("spectrum") and Piroda ("nature"), are planned to be added in the late 1990s.

Missions to Mir

The first flight to Mir was launched on March 13, 1986; it docked with Mir two days later. Crew members Leonid Kizim and Vladimir Solovyev flew aboard the Soyuz T15. Upon reaching Mir, the cosmonauts activated its systems and received and unloaded supplies that had arrived in two unpiloted Progress vehicles. They made films of Earth, established a communications link with the ground using the geostationary satellite Lyutch, and performed several tests using the space station's computers.

Cosmonauts aboard the Mir space station demonstrate some of the effects of living and working in weightlessness. The floor, ceiling, and walls in the work area of Mir are painted different colors to help the crew keep track of directions.

After almost two months on Mir, Kizim and Solovyev flew the Soyuz T15 to the empty, orbiting Salyut 7 space station, where they spent 50 days. This was the first space transfer from one station to another. They returned to Mir, bringing the results of experiments, films, photographs, and equipment from Salyut 7. After spending three more weeks on Mir, they returned to Earth and ended their 125-day mission.

The next cosmonauts to visit Mir were Yuri Romanenko and Aleksandr Laveïkin in the Soyuz TM2, launched in February

1987. After unloading the Progress 27 cargo vehicle, they set up a satellite relay link and prepared Mir to receive the Kvant module. This was planned as the world's first yearlong mission, during which there would be three visits by other cosmonauts. Romanenko was in space for 326 days and readapted to Earth life well, confirming that long-duration flights were possible. Laveïkin, however, suffered from health problems aboard Mir and was replaced by Aleksandr Aleksandrov.

On December 23, 1987, Vladimir Titov and Musa Manarov replaced Romanenko and Aleksandrov, who returned to Earth on December 29, with Anatoli Levchenko piloting their return. Titov and Manarov stayed on Mir for a year, conducting astrophysical, technological, medical, and biological research. Three crews visited during their stay.

Sergei Krikalev and Anatoli Artsebarskii arrived at Mir in May 1991. Artsebarskii returned in October, but Krikalev's return trip was delayed for several months until March 25, 1992 due to changes prompted by the breakup of the Soviet Union.

In 1993 Russia and the United States scheduled future joint missions involving Mir and the U.S. Space Shuttle. These will draw on decades of Russian expertise in space station operations and the proven capabilities of the U.S. Space Shuttle.

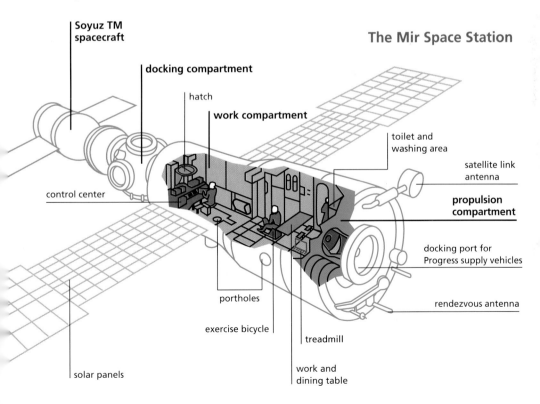

The Mir Space Station

Soyuz TM spacecraft

docking compartment

hatch

work compartment

toilet and washing area

satellite link antenna

control center

propulsion compartment

docking port for Progress supply vehicles

rendezvous antenna

portholes

exercise bicycle

treadmill

solar panels

work and dining table

elevons (combined ele-
vators and ailerons)
that control flight on
return to atmosphere

Orbital
Maneuvering
System engine

main engines used
during launch

rudder and speed
brake used in landing
located on tail

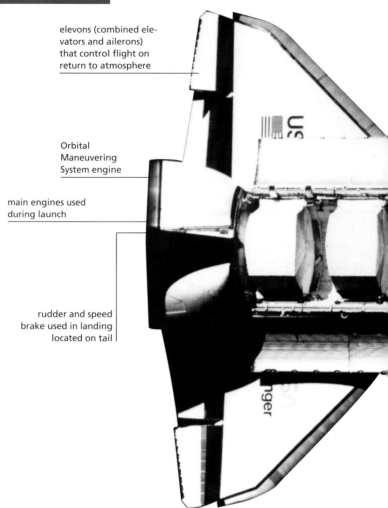

The Space Shuttle *Atlantis*
lifts off in October 1989,
carrying a crew of five
and the spacecraft
Galileo, which was then
deployed from the
Orbiter and set on course
to carry a probe to the
planet Jupiter—a journey
expected to take more
than six years.

The U.S. Space Shuttle Program

The American Space Shuttle, officially called the Space Transportation System (STS), is the only reusable spacecraft operating today.

Once the Moon race had been won, NASA began seeking congressional approval of a space station and an STS by promoting the Space Shuttle's use as a ferry to and from a space station and as a scientific laboratory; a means of repairing, resupplying, or recovering satellites; and a carrier of commercial payloads such as satellites. Another major selling point was that the crews could return to Earth with valuable equipment.

The project received congressional approval and was announced to the American public early in 1972. The first flight in the series occurred in 1981.

Space Shuttle Orbiter

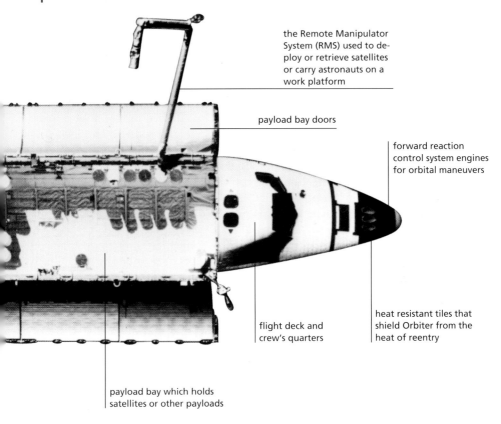

the Remote Manipulator System (RMS) used to deploy or retrieve satellites or carry astronauts on a work platform

payload bay doors

forward reaction control system engines for orbital maneuvers

flight deck and crew's quarters

heat resistant tiles that shield Orbiter from the heat of reentry

payload bay which holds satellites or other payloads

Shuttle Components

The Space Shuttle comprises three main parts: the Orbiter, the Solid Rocket Boosters (SRBs), and the External Tank. The Orbiter looks much like an airplane and is about the size of a DC-9 jet. Total height is about 184 feet (56 m), about half that of the tall, slim Saturn V rocket. Wingspan is about 78 feet (24 m). With the exception of its windows, the entire Orbiter is covered with insulation to protect it from the heat generated during reentry. On the underside, where reentry heat is extreme, there are about 23,000 individual ceramic-coated, silica fiber tiles that prevent the Space Shuttle Orbiter from being incinerated over the course of dozens of reentries.

Inside the Orbiter Astronauts reside and work in a two-level crew compartment in the forward fuselage of the Orbiter, which can accommodate up to eight people (ten in an emergency). The upper level houses the flight deck "cockpit" and a workstation, which controls the payload bay operations. The mid-deck is used for sleeping, eating, storage, and small experiments; it also has an airlock that opens into the payload bay and allows access to space. A lower deck below the mid-deck floor is the site of equipment such as water pumps and air purification systems. In the center of the craft is a payload bay 60 feet (18.3 m) long and 15 feet (4.6 m) wide—about the size of a tour bus. Up to 65,000 pounds (29,500 kg) of payload can be lifted into a low orbit; 32,000 pounds (14,500 kg), into a polar orbit. The Remote Manipulator System (RMS) is a mechanical arm that moves objects in and out of the payload bay and can mount a work platform for astronauts. Two large payload bay doors are opened to deploy satellites or to expose scientific instruments to space. The doors also have radiators to dispose of heat produced by the electrical equipment and the astronauts; these doors must be opened on flights to prevent the shuttle from overheating. Electrical energy on shuttle flights is provided by fuel cells that use oxygen and hydrogen.

The Space Shuttle's guidance, navigation, and control depend on its five general purpose computers. The shuttle's autopilot is one of the most advanced ever built, reacting many thousands of times faster than any human being. One of the computers acts as a backup during launch and reentry—the two most critical phases of the mission. The reliability of the other four computers is extremely high—during critical phases, they serve as backup for one another.

In preparation for a Space Shuttle mission, this huge external propellant tank, which holds liquid oxygen and liquid hydrogen, is on its way to the Kennedy Space Center's Vehicle Assembly Building.

Typical Space Shuttle Flights

Unlike the rockets before them, which lifted off slowly and seemingly reluctantly, the Space Shuttle leaves the launchpad within three seconds after ignition. A thrust of almost 6 million pounds (2.7 million kg) is delivered by the solid rockets, with 1.1 million additional pounds (.49 million kg) created by the main engines. And while earlier launches subjected the crew to higher g forces, shuttle crews experience peak acceleration forces of 3 g only twice during a mission. (This force is actually less than you experience on some roller coasters.) The first occurs about two minutes after liftoff, when the Solid Rocket Boosters burn out and fall away. The second occurs about 8 minutes into the flight, just before the large External Tank separates.

An Orbiter can remain in space for up to 14 days. When its mission in space ends, the Orbital Maneuvering System engines are fired to reduce the Orbiter's speed, and the shuttle reenters the atmosphere. As the air becomes thicker, aerodynamic controls become effective. Though the shuttle is unpowered at this point, it can fly for several hundred miles and land much as a glider lands. The Orbiter is then refurbished for the next flight.

Of the five Orbiters produced, all but *Challenger,* which was destroyed in an explosion in 1986, are in the present fleet. *Columbia* was first flown in 1981; *Challenger,* in 1983; *Discovery,* in 1984; *Atlantis,* in 1985; and *Endeavour* in 1992.

Space Shuttle Cockpit

environmental controls

fuel cell controls

altitude/vertical velocity

attitude indicator

commander's position

compass

flight computer

autopilot controls

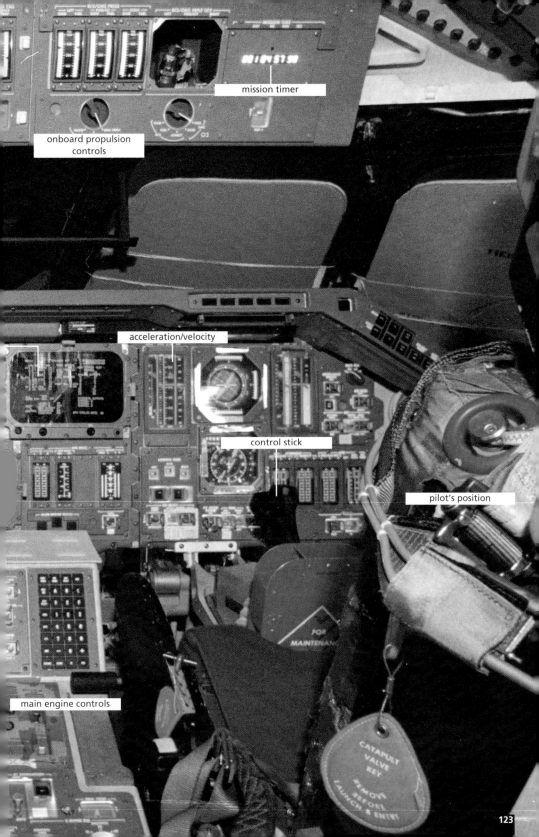

mission timer

onboard propulsion
controls

acceleration/velocity

control stick

pilot's position

main engine controls

Astronaut Jerry L. Ross performs extravehicular activities (EVA) during a Space Shuttle *Atlantis* mission in 1985. The Remote Manipulator System (RMS) arm moves Ross to his work site. He is secured by a foot restraint that keeps him from floating away into space.

Physical exercise is an integral part of Space Shuttle missions. During a 14-day mission aboard *Columbia* in 1992, astronaut Ellen S. Baker works out on the bicycle ergometer, located mid-deck.

Everyday Life on the Shuttle

Shuttle crews, including NASA astronauts and payload specialists (researchers drawn from government, industry, and universities), accomplish a variety of scheduled tasks during their missions. In addition to basic Orbiter operations and housekeeping duties, work by crews includes conducting research and experiments, deploying and retrieving satellites, performing in-orbit repairs, taking photographs, and doing TV and film work.

Experiments are conducted by scientists from many nations in Spacelab, a modular laboratory built by the European Space Agency. Carried in the payload bay of the Orbiter, the laboratory is exposed to space when the payload bay doors are opened.

The crew's demanding schedules call for long work days, but they must also exercise to counteract the effects of weightlessness. Members of the shuttle team regularly use a small treadmill and a variety of bungee cord devices.

Daily menus are designed to restore the calories and minerals expended during strenuous space duties. The selection and quality of food on shuttle flights have improved greatly since John Glenn first squeezed a tube of applesauce in his mouth during his orbital Mercury flight in 1962. Shuttle crews now have a choice of over one hundred food items. Varied meals can be enjoyed for six days before a meal is repeated.

Shuttle Mission Highlights

As of July, 1994, 62 shuttle missions have been flown, with three more scheduled in 1994. Each mission has its own specific goals and objectives.

Enterprise Before *Columbia* went into orbit in 1981, a test shuttle named *Enterprise* was flown to an altitude of 22,000 feet (6,700 m) atop a Boeing 747 jet in 1977. It was then released from the 747 to perform approach and landing tests in Earth's atmosphere. Engineers wanted to confirm wind tunnel tests and to be certain that the 150,000-pound (68,040 kg) vehicle could land safely. *Enterprise,* the heaviest glider ever flown, was piloted to a landing strip at Edwards Air Force Base and landed successfully on each test flight.

Columbia: Mission STS-1 On April 12, 1981, *Columbia* was launched as an orbital flight test, with commander John Young and pilot Robert Crippen aboard. This was the first time that a shuttle had been piloted, operational, and in orbit, so there were many potential risks. Only the gliding portion of the flight had been tested. No payload was carried on this test mission. The crew successfully completed orbital maneuvers and many other operational tests. After ten years in development, *Columbia* was performing as designed.

The descent through the atmosphere created temperatures of about 2,750° F (1,510° C), which the insulating outer tiles withstood, as planned. After the expected 16-minute communications blackout while *Columbia* reentered the atmosphere, radio contact was again established. Mission control informed Young and Crippen that all was well. Minutes later, after a 54-hour flight making 36 Earth orbits, *Columbia* touched down. Four more *Columbia* flights were completed before the second Space Shuttle, *Challenger*, was launched.

Astronaut Dale A. Gardner (left), wearing the Manned Maneuvering Unit (MMU), approaches the spinning Westar VI satellite over Bahama Banks, during the November 1984 Space Shuttle *Discovery* mission. After repairing a satellite, astronaut James D. van Hoften (right) conducts his first in-space field test of the MMU in the payload bay of *Challenger* in April 1984.

"The dream is alive again."
—John Young aboard the first
Space Shuttle, *Columbia*

Challenger: Mission STS-6 By the time *Challenger* was first launched on April 4, 1983, all shuttle systems had been tested in actual flight. This flight, however, carried one of the largest payloads ever: 46,615 pounds (21,144 kg), including a tracking and data relay satellite and the upper stage booster that would take the satellite into geosynchronous orbit. The satellite was the first of six to be launched to create a network for continuous tracking.

To accommodate the extra weight, design changes were made to the *Challenger* spacecraft, including making the external tank and solid rocket boosters 20,000 pounds (9,070 kg) lighter. This flight included a space walk during which the astronauts practiced repair techniques in the open payload bay. The mission ended after five days and a perfect landing.

Included in the crew on *Challenger*'s second flight in June 1983 (STS-7) was Sally K. Ride, the first U.S. woman astronaut in space.

This low-angle view of the Space Shuttle *Discovery* was taken as the spacecraft prepared for landing in 1990. It had completed a successful five-day Earth orbital flight, during which the Hubble Space Telescope (HST) was launched.

Discovery The third Space Shuttle, and the 12th shuttle flight in the program, lifted off on August 30, 1984, with a crew of six. The crew included Judith Resnik, the second U.S. woman astronaut in space, as well as the first commercial payload specialist, a scientist from McDonnell Douglas. During the flight, three communications satellites were launched from the payload bay toward geosynchronous orbit, and solar array panels were deployed and tested. This first *Discovery* flight lasted six days, during which 97 orbits were completed.

Atlantis *Atlantis* was the fourth spacecraft introduced in the shuttle fleet, and its first mission was the 21st shuttle flight. It was launched on October 3, 1985, on a classified military mission for the U.S. Air Force. The purpose of the mission was believed to be the deployment of two defense communications satellites. *Atlantis* is being modified to prepare it for missions to the Mir space station in the mid to late 1990s.

The Final Challenger Flight The *Challenger* was scheduled to make the 25th shuttle flight on January 28, 1986. There were seven crew members—Richard Scobee, Michael Smith, Ronald McNair, Ellison Onizuka, Judith Resnick, Gregory Jarvis, and civilian Christa McAuliffe, a teacher. Less than two minutes into the flight, however, the propellants exploded and the vehicle broke apart. The entire crew died in the accident. A leak in a joint in one of the Solid Rocket Boosters was the cause. The shuttle program was suspended for two years while the joint was redesigned and safety precautions and decision-making procedures were changed. Never again would any shuttle launch be considered "routine."

Astronauts (below left) Robert L. Gibson, the mission commander, and pilot Curtis L. Brown Jr., wear partial-pressure launch and entry suits as they power down the Space Shuttle *Endeavour* to end its seven-day Spacelab-J mission, September 1992. Crew members (bottom right) disembark from the Space Shuttle *Challenger* following a five-day mission completed on April 9, 1983. Commander Paul Weitz leads the group, followed by mission specialists Story Musgrave and Donald Peterson, and pilot Karol Bobko.

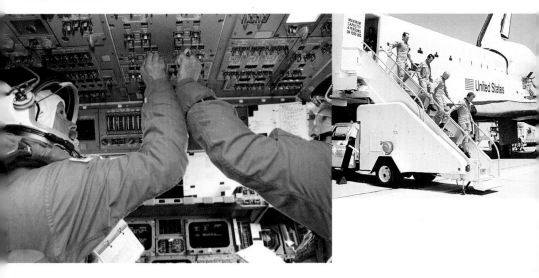

Space Suits and Extravehicular Activity (EVA)

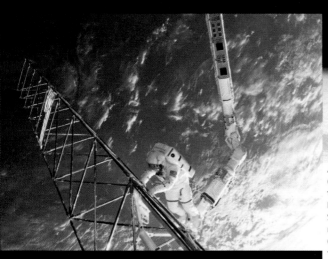

Astronaut Jerry L. Ross (left) is anchored to the foot restraint of the RMS on an *Atlantis* Shuttle mission. Astronaut Bruce McCandless (top) takes the world's first untethered space walk as he tests the MMU for the first time in February 1984.

In the early days of piloted spaceflight, pressurized suits contained as many as 15 layers. Even moving the fingers was difficult. Gradually, astronauts' space suits were redesigned to increase mobility, and comfort. The air pressure in the Space Shuttle and other piloted spacecraft in use today is equal to Earth's, and daily wear is usually layers of lightweight cotton clothing. Astronauts only don space suits during critical phases of the mission or for EVA (Extravehicular Activity).

During EVA, U.S. astronauts use an Extravehicular Mobility Unit (EMU), which includes a space suit, a life support system, and a command system. For special EVA projects, such as inspecting a satellite, astronauts use the Manned Maneuvering Unit (MMU) to aid mobility.

The space suit used on the Space Shuttle consists of 11 layers of Nylon, Dacron, and Kevlar, plus an outer fiberglass shell for the torso, and weighs about 104 pounds (4 kg) on Earth. The helmet weighs 8 pounds (47 kg). The life support system attached to the back of the upper part of the space suit

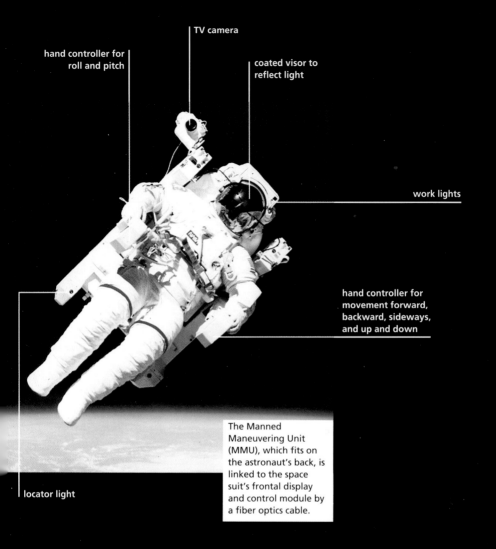

TV camera

hand controller for
roll and pitch

coated visor to
reflect light

work lights

hand controller for
movement forward,
backward, sideways,
and up and down

The Manned
Maneuvering Unit
(MMU), which fits on
the astronaut's back, is
linked to the space
suit's frontal display
and control module by
a fiber optics cable.

locator light

weighs 148 pounds (68 kg). It contains oxygen, rechargeable batteries, an air treatment system, and water for a cooling system—all of which enable an astronaut to remain outside the shuttle for up to eight hours with a small reserve. The display and command system on the front of the space suit allows the astronaut to monitor the life support system and to communicate with the crew inside the spacecraft.

The MMU enables an astronaut to move about in space without being tethered to the spacecraft. It contains maneuvering jets powered by compressed nitrogen. Hand controls, similar to those on an interactive video game, allow the astronaut to move in any direction.

During Space Shuttle EVAs, astronauts may also be attached to the Remote Manipulator System (RMS)—a mechanical arm operated from inside the shuttle that moves objects in and out of the payload bay in space. The RMS has three joints: shoulder, elbow, and wrist. It also has two television cameras on it to allow observation of the "arm" movements.

In order to move Space Shuttles cross-country, NASA uses a specially modified Boeing 747. Here, on its way to Houston for a July Fourth ceremony in 1982, *Challenger* flies piggyback on NASA's shuttle carrier—a spectacular sight from air or land.

Endeavour The newest of the shuttles, *Endeavour* was built to accommodate 16-day missions. The accomplishments of the first five *Endeavour* missions include experiments in material processing and biotechnology and space walks to service the Hubble Space Telescope. The STS-49 mission in May 1992—*Endeavour*'s first flight—retrieved and reboosted the communication satellite Intelsat VI. On its sixth mission in April 1994, the 62nd Space Shuttle flight, *Endeavour*'s Space Radar Laboratory payload provided the international scientific community with vital environmental data, including facts about pollution and global distribution of carbon monoxide.

The Soviet Shuttle Buran

The dimensions and layout of *Buran* ("blizzard" or "snowstorm") and the U.S. shuttles are basically the same; both have a payload bay and a two-level crew compartment. Both vehicles are designed to be reusable and to carry passengers. Unlike the U.S. Space Shuttle, however, *Buran* does not have to be piloted; it can be flown and landed either on autopilot or by two pilots. *Buran* does not have its own launch system. It is launched by the Energiia, which releases the shuttle just before it reaches orbital altitude, after which small rockets propel it into orbit.

Buran is designed to carry 30 tons (27 MT) of payload at launch and return as much as 20 tons (18 MT) to Earth. The propulsion system is located in the rear compartment. It has two

The *Buran* shuttle, mounted on the Energiia booster, is transported to the launchpad for its November 1988 mission, its only flight to date. As the shuttle's future is uncertain, Russia has put surplus *Buran* airframes up for sale on the open market.

main rocket engines, 38 attitude control thrusters, and eight precision thrusters. Fine control in orbit is provided by two rings of engines, the first at the front of the fuselage and the other at the back of the tail compartment. More than 50 subsystems make up the control system, and all are controlled by four computers that could activate a reserve subsystem should any fail.

Reportedly, the Soviet space shuttle can remain in constant contact with Earth through relay satellites. *Buran* is claimed to be capable of launching satellites, even from unpiloted flights, and returning satellites to Earth.

Buran's initial, and thus far only, flight on November 15, 1988, was unpiloted. It lasted three hours and 25 minutes and orbited Earth twice. During this test flight, engineers evaluated the Energiia launch, *Buran's* entry into orbit, and its descent, reentry, and landing. Although this test flight was a success, the *Buran* program has been put on hold due to political and economic changes in the former Soviet Union.

Exploring the Moon

Destination Moon: the hauntingly beautiful natural Earth satellite and inspiration of writers and dreamers for centuries. The myriads of technical obstacles to piloted flight were indeed daunting. On July 20, 1969, however, the first humans were poised to land on the lunar landscape, and when this staggering achievement occurred, it enthralled the world. As Wernher von Braun expressed it: "I think it is equal in importance to that moment in evolution when aquatic life came crawling up on the land." At the very least, thoughts of Moon gazers worldwide had been unalterably changed.

Beautiful yet desolate images, such as this one of a lunar crater, were captured during the Apollo 15 missions.

Getting Started

On May 25, 1961—more than three years after the stunning launch of the Soviet Sputnik satellite—President Kennedy dramatically directed NASA to put a human on the Moon by the end of the 1960s:

> "I believe that this nation should commit itself to achieving the goal, before this decade is out, of landing a man on the Moon and returning him safely to the Earth."

With this challenge, the United States plunged into the Space Race with the Soviet Union.

Soviet Moon Probes

Luna and the Lunokhod In 1958, the Soviets launched a series of robotic Moon probes named Luna. In January 1959 the first probe, Luna 1, missed the Moon; but in September, Luna 2 became the first human-made object to hit the Moon. In October, Luna 3 photographed much of the far side of the Moon when its highly elliptical Earth orbit passed behind the Moon. Launched on January 31, 1966, the more complex Luna 9 became the first probe to soft-land and relay TV pictures of the lunar surface to Earth on February 3. On two Luna missions in 1970 and 1973, a robotic roving vehicle named Lunokhod was deployed to study soil and rock samples by drilling, and to take photographs. In addition to the two rover missions, the Soviets also completed three successful robotic sample return missions in 1970, 1972, and 1976. Although these missions returned far less lunar soil than rival U.S. missions, they did demonstrate that successful robotic exploration was possible.

Lunokhod's powerful digging mechanisms were able to collect lunar soil samples from well below the Moon's surface. The Lunokhod vehicles were employed during the missions of the robot Luna probes.

Zond The Soviets also launched a series of eight Moon probes named Zond from 1964 through 1970. The Zond missions were test flights, precursors to a piloted lunar mission. They took close-up pictures of much of the lunar surface, including the far side. In 1968 Zond 5 became the first spacecraft to orbit the Moon and return to Earth.

U.S. Moon Probes

Pioneer The Pioneer lunar probes (1–7, launched from 1958 to 1960) and Ranger probes that followed were designed to take photos from space and then crash-land on the lunar surface. Although only one probe in the early Pioneer series, Pioneer 4, was successfully launched into space, valuable data resulted regarding radiation levels in space and better spacecraft guidance.

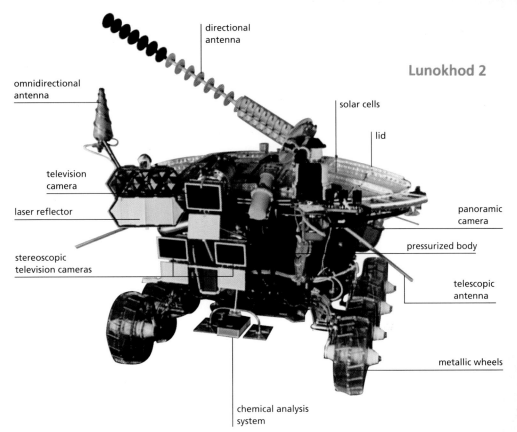

directional antenna

omnidirectional antenna

solar cells

lid

television camera

laser reflector

panoramic camera

pressurized body

stereoscopic television cameras

telescopic antenna

metallic wheels

chemical analysis system

Ranger The U.S. Ranger series consisted of nine spacecraft launched intermittently between 1961 and 1965. The first six failed, but the last three (Rangers 7, 8, and 9) returned high-quality images of the Moon—providing a first close look at the lunar surface. NASA data was impressive but left open the question of whether or not the surface of the Moon could support a landing of a spacecraft and crew.

Surveyor Five of the seven U.S. Surveyor probes proved successful. Surveyor 1, for example, soft-landed on the Moon on June 2, 1966, and the quality of the data transmitted by subsequent Surveyor missions continued to improve. In 1967 Surveyor 6 lifted off the lunar surface, moved to a new location, and set down to photograph another area—the first time an engine had been restarted on an extraterrestrial body. Surveyor data gave assurance that the lunar soil could support a spacecraft, and that astronauts would be able to walk on the Moon's surface.

Technicians test a Surveyor probe in California in 1966.

Lunar Orbiters In August 1966, NASA launched a series of five lunar orbiters. They relayed high-resolution images of a vast area of the lunar surface, and a potential landing site was selected for future astronaut missions—the Sea of Tranquility.

In 1962, speaking at Rice University, President John F. Kennedy outlined America's goals in space and inaugurated the decade of the Apollo space program. In preparation for the July, 1971 launch, Apollo 15 astronauts David Scott and James Irwin (right) test drive a Lunar Roving Vehicle on a simulated lunar surface in Taos, New Mexico.

Apollo: Destination Moon

In preparation for landing humans on the Moon—what has been called one of the most difficult technological efforts humans have ever undertaken—NASA captured the imagination of the world. The Moon has inspired worship and legends since ancient times. Although "moonlight" is merely the reflection of light from the Sun, the Moon is heralded as the brightest wonder of the skies. Although the lunar atmosphere lacks air, wind, or water and the sky is always black there, popular songs, such as "Fly Me to the Moon" and "By the Light of the Silvery Moon," help maintain the idea that this satellite, which does not sustain life, would be a wondrous, romantic place to visit.

Launch and Flight Hardware

NASA became partners with industry to develop rocket engines for the lunar launch vehicle. North American Rockwell Corporation was awarded the major contract for the Command Module (CM) and Service Module (SM). Based on the design of the Mercury and Gemini capsules, the CM was expanded to accommodate three astronauts. The Grumman Aircraft Engineering Corporation won a contract to design the Lunar Module (LM), which would land on the Moon with two astronauts in it. Space suits, backpacks the astronauts would wear while on the surface of the Moon, and all the other Apollo necessities also had to be designed and produced by a huge contractor team.

Apollo 13 commander, astronaut James A. Lovell Jr., sets down a Lunar Landing Training Vehicle at the conclusion of a test flight for the planned mission. A hovering helicopter crew supervises Lovell's landing.

Apollo Astronaut Training

While NASA research centers were studying the structure and control of launch vehicles and spacecraft, astronauts were being recruited and trained. Seven were selected in 1959, nine in 1962, and fourteen in 1963. The first groups of recruits were military test pilots with engineering degrees. In 1965 the six astronauts selected marked a departure from earlier groups—they were the first group of scientists. In 1966, another nineteen pilots were accepted for astronaut training. Training for the Moon-bound astronauts was both extensive and intensive. Expanding on the training of the Mercury and Gemini astronauts, they studied celestial and lunar navigation, did survival training in deserts and rain forests, visited areas thought to be like lunar surfaces, experienced simulated lunar activity in a one-sixth gravity trainer, and practiced water recovery routines.

Once the design of the LM (in which two astronauts would land on the Moon's surface) and the CM (in which one astronaut would orbit the Moon) was fairly well established, simulators for both were built, and the astronauts began detailed training. Each crew was required to spend at least 140 hours in the LM simulator and 180 hours in the CM simulator before they were considered qualified for flight. To be prepared for either a malfunction or an unexpected occurrence in space, the crew had to know the intricacies of the systems in both modules and be familiar with their backup systems.

To learn about hazards in the near-Earth space environment, three Pegasus micrometeoroid detection satellites were launched in 1965. These satellites, which carried huge panels nearly 100 feet (30 m) long , determined that the rate of meteoroid penetrations was much lower than anticipated. This finding influenced Apollo hardware design and allowed a reduction in the overall weight of the Apollo spacecraft of about 1,000 pounds (454 kg).

The official design for the Apollo program patch consists of the letter A (its crossbar made up of the three central stars in the constellation Orion), the Earth, and the Moon. The face on the Moon represents the mythological god Apollo. A double trajectory passes behind both spheres and through the central stars.

The crawlerway from the Vehicle Assembly Building to the launch complex is only 3.5 miles (5.6 km) long. However, it seemed endless during the one mile per hour (1.6 km/hr) transport of the Saturn V-Apollo space vehicle.

Mission Control

Four years and $800 million were required for construction of the needed lunar mission facilities in Texas and Florida.

In Houston Many NASA centers contributed to the Apollo project, headquartered in Houston, Texas. Facilities for the design, testing, and inspection of piloted flight elements, as well as medical facilities for the care, monitoring, and postflight quarantine of astronauts, were built at the Manned Spacecraft Center in Houston. Training facilities for astronauts were also constructed there, including all simulators. Houston's mission control facilities monitored the spacecraft once it cleared the launch tower—well before it was in space.

At Cape Canaveral The Vehicle Assembly Building (VAB) was constructed at the Cape Canaveral Space Center in Florida (now known as the Kennedy Space Center). The VAB could house four of the Saturn-Apollo launch vehicle spacecraft assemblies, each standing 36 stories high. Work platforms at various levels gave access to all components of the Saturn V during assembly. Two launch pads were constructed, with very level and stable roads leading from the VAB to the pads.

A vehicle dubbed a "crawler," designed and built with a tank-like tractor unit at each corner, transported a complete vertical Saturn-Apollo assembly from the VAB to the pads, moving at the cautious rate of one mile per hour (1.6 km/hr). These crawlers were the world's largest land vehicles.

A New View of the Moon

While the Moon makes one orbit around Earth, it also makes exactly one rotation on its axis. This means that the same side is always turned toward Earth. No one had seen the far side of the Moon until the Soviet Union's Luna 3 photographed it in 1959 and transmitted the pictures to Earth.

Apollo 17 astronauts photographed this spectacular view of a full Moon, one-third of which shows the lunar far side, during the final Apollo mission.

Each launchpad was equipped with a 45-story gantry, a scaffolding that allowed servicing and fueling of the vehicle on the pad, as well as access to the CM for the astronauts. A launch control center was also constructed, from which the operation of the first two launch stages was controlled before transferring control to Houston.

To maintain deep space tracking for lunar and planetary spacecraft, three powerful antennas were placed equidistantly around the globe—at Goldstone, California; near Madrid, Spain; and near Canberra, Australia. Used initially for Mercury and Gemini missions, these large antennas were also essential elements for tracking the Apollo missions.

Unpiloted Launches

In preparation for piloted flight, several unpiloted launch tests, mostly involving the Saturn I rocket and the variant Saturn IB, were completed from 1961 to 1966. Saturn IB, for example, proved capable of placing 40,000 pounds (18,160 kg) into low Earth orbit. Five tests were made to launch models of the Apollo CM. Finally, on November 9, 1967, the much larger Saturn V launch vehicle, which would become the primary workhorse of the Apollo program, was test-fired at NASA's Mississippi facility on the Pearl River. It proved capable of launching 285,000 pounds (129,390 kg) into Earth orbit, or 100,000 pounds (45,400 kg) to the Moon.

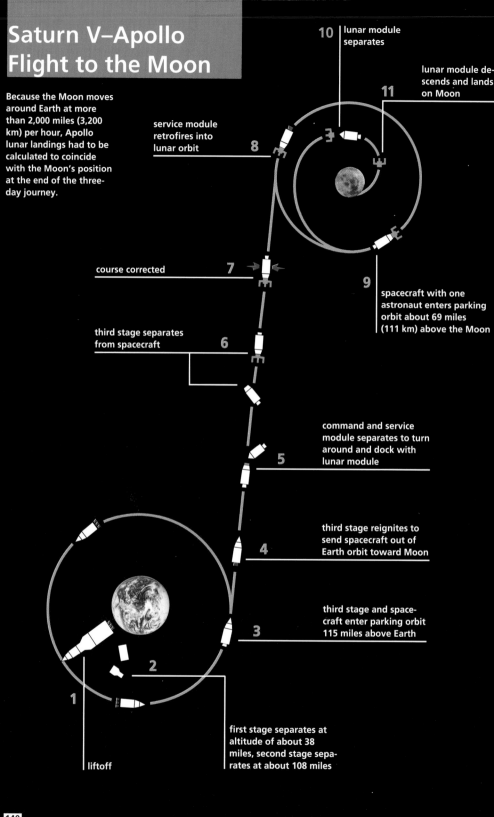

Saturn V–Apollo Flight to the Moon

Because the Moon moves around Earth at more than 2,000 miles (3,200 km) per hour, Apollo lunar landings had to be calculated to coincide with the Moon's position at the end of the three-day journey.

10 lunar module separates

lunar module descends and lands on Moon

11

service module retrofires into lunar orbit **8**

course corrected **7**

9 spacecraft with one astronaut enters parking orbit about 69 miles (111 km) above the Moon

third stage separates from spacecraft **6**

command and service module separates to turn around and dock with lunar module **5**

third stage reignites to send spacecraft out of Earth orbit toward Moon **4**

third stage and spacecraft enter parking orbit 115 miles above Earth **3**

2

1

first stage separates at altitude of about 38 miles, second stage separates at about 108 miles

liftoff

The Saturn V–Apollo

Beginning with Apollo 8 in December 1968, the Saturn IB was officially replaced by the much larger, more powerful Saturn V. The first stage alone of the three-stage Saturn V was as tall as an entire Saturn IB. Each of the five engines of Saturn V's first stage had the power of all eight engines combined on the Saturn IB used on Apollo 7.

Above the three rocket stages was a short section containing the partly folded Lunar Module (LM). Then came a larger section containing the Service Module (SM) with the nozzle of its main engine protruding from the rear of the module. In addition to this engine, the SM also had four groups of four thrusters designed to maneuver the spacecraft.

Above this was the cone-shaped Command Module (CM), with the control cockpit designed to hold three astronauts. The CM contained 24 display instruments, 40 mechanical event indicators, 71 lights, and over 560 switches. The large flat end of the CM was protected with a heat shield. The SM and CM were attached and collectively known as the CSM.

The CSM was designed to separate from the LM after launch, turn 180 degrees, and dock with the LM, now attached to the nose of the CM module for the trip to the Moon. The SM, however, was to jettison just before reentry to Earth; the CM was designed to return to Earth safely and splash down in the ocean. At the top of the stacked modules was the escape rocket, designed to be used in an emergency to pull the CM away from the rest of the vehicle, either on the pad or during launch.

The Apollo 15 Command and Service Module is shown before and during the mission. The CSM (right) is seen in lunar orbit hovering above the Moon's Sea of Fertility during rendezvous with the Lunar Module.

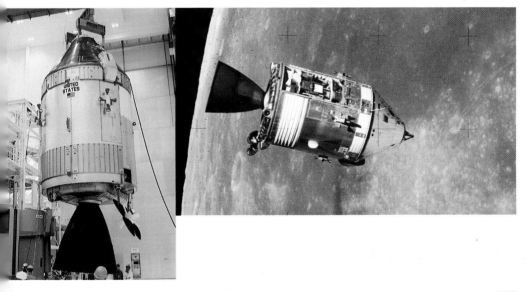

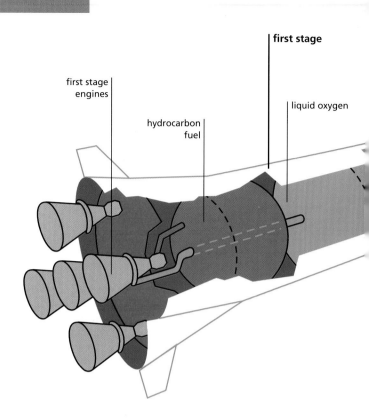

first stage

first stage
engines

hydrocarbon
fuel

liquid oxygen

The Saturn V-Apollo 11 space vehicle is carefully inched down the crawler-way in preparation for the historic lunar landing mission—the first human footsteps on the Moon—in July, 1969.

Apollo Missions: To the Moon and Back

Twelve piloted Apollo launches and three unpiloted launches preoccupied NASA personnel in the years 1967 to 1972. During those years, the lunar program that began with Apollo 1 and ended with Apollo 17 was not without its tragedies and setbacks. The United States could, however, find cause for celebration in the daunting technological challenges that were met to achieve its space objectives. Apollo missions 18, 19, and 20 were cancelled as unnecessary—which represented not only a tribute to NASA's amazing successes but also a decline in public interest and funding during a recession-prone era.

Apollo 1 Apollo 1 was scheduled to liftoff on February 21, 1967. The selected crew consisted of Virgil "Gus" Grissom, Roger Chaffee, and Edward White. In a training exercise on the launchpad on January 27, the three astronauts had been sealed in the CM for over five hours when a fire started by an electrical malfunction broke out in the cockpit. The atmosphere in the spacecraft was pure oxygen, so the spark immediately became an inferno. The crew were quickly asphyxiated. After this tragedy, the CM was redesigned to include a quick-escape hatch, and to have a mixed-gas cabin atmosphere rather than pure oxygen.

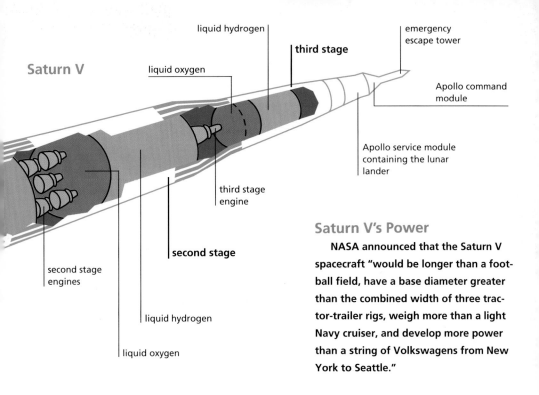

Saturn V

liquid hydrogen

liquid oxygen

third stage

emergency escape tower

Apollo command module

Apollo service module containing the lunar lander

third stage engine

second stage

second stage engines

liquid hydrogen

liquid oxygen

Saturn V's Power

NASA announced that the Saturn V spacecraft "would be longer than a football field, have a base diameter greater than the combined width of three tractor-trailer rigs, weigh more than a light Navy cruiser, and develop more power than a string of Volkswagens from New York to Seattle."

Apollo 4, 5, and 6 were unpiloted test flights flown in 1967 and 1968. Apollo 4 was the first test launch of the Saturn V, which generated 7.5 million pounds (3.4 million kg) of thrust. Apollo 5 was the first test of the unpiloted Lunar Module in Earth orbit during which the LM's ascent and descent engines were fired successfully twice.

The launch of Apollo 6 on April 4, 1968, was an unpiloted dress rehearsal. But while it succeeded in putting Apollo hardware into Earth orbit, there were fuel flow problems and engine failures in the Saturn launcher. Each of the next Apollo missions, however, brought the program nearer to its goal as they approached ever closer to the Moon's surface and practiced the coordinated maneuvers between the LMs and CMs that would be needed for an eventual lunar landing.

Apollo 7 October 11–22, 1968. Crew: Walter M. "Wally" Schirra Jr., Donn F. Eisele, and R. Walter Cunningham

Launched by the Saturn IB, Apollo 7 was the first piloted launch of Apollo—the engineering shakedown flight to test all hardware and systems. For almost eleven days during 163 Earth orbits, the three astronauts (plagued with head colds during the mission) tested guidance and control systems, a new space suit design, food supplies, and the work-rest-sleep routine. They also tested the restarting capability of the CSM rocket and the simulated rendezvous of the CSM with an LM.

Lunar Module

The Lunar Module (LM) was designed for the low-gravity, no-atmosphere environment of the Moon. It was an ungainly and flimsy-looking structure, not like the more substantial, streamlined CM. LMs were not brought back to Earth. The robotic-looking LM had two windows, a hatch for entry and exit, and four spindly legs that earned it the nickname "Spider." The LM—with self-contained computer, guidance, and propulsion systems—was a feat of technology and lightweight engineering.

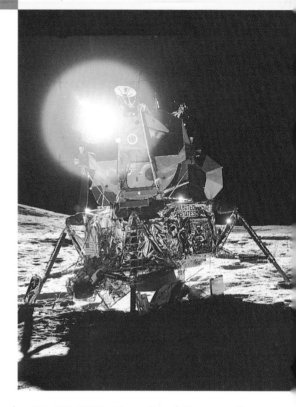

Apollo 8 December 21–27, 1968. Crew: Frank Borman, James A. Lovell Jr., and William A. Anders

Apollo 8, the first human lunar orbital flight, was the most important space achievement leading to the Moon landing by Apollo 11. Two months after the return of Apollo 7, the Saturn V launch vehicle was used for the first time on a piloted flight. The voyage covered about 500,000 miles (804,700 km), including ten orbits of the Moon over a period of 16 hours. The total time of the mission was 147 hours (just over six days).

The astronauts on Apollo 8 were the first people to view the entire Earth from space and the first to see the back side of the Moon. Apollo 8 was also the first piloted flight to escape Earth's gravitational field and be influenced by the Moon's gravity.

The astronauts transmitted stirring observations from lunar orbit, describing the desolate lunar landscape that made Earth seem, by comparison, "a grand oasis" and reading passages from the Book of Genesis on Christmas Eve, 1968.

Apollo 8 emerged from the far side of the Moon for the last time as Lovell announced the successful firing of its reentry propulsion system—the only way it could return to Earth—with the words: "Please be informed there is a Santa Claus." Splashdown in the Pacific was successfully executed on December 27. Although Soviet probes Zond 5 and Zond 6 of September and

Astronaut David R. Scott performs a stand-up extravehicular activity (EVA) on the fourth day of the Apollo 9 Earth-orbital mission. Scott, the command module pilot, is standing in the open hatch of the CM.

November 1968 respectively had suggested that the Soviets were close to a piloted mission, Apollo 8 proved that the United States was clearly the leader in space.

Apollo 9 March 3–13, 1969. Crew: James A. McDivitt, David R. Scott, and Russell L. "Rusty" Schweickart aboard *Gumdrop* (CM) and *Spider* (LM)

The Apollo 9 astronauts accomplished the first piloted test flight of the LM, including docking, while in Earth orbit. At one time the LM, carrying McDivitt and Schweickart, was over 100 miles (161 km) away from the CSM, too far away to be seen. The LM rendezvous and docking were controlled by the commander in the LM. This mission also included multiple firings of the SM engine. Apollo 9 landed after 151 orbits. The LM had flown separately from the main spacecraft for six hours. Another Apollo 9 achievement was a space walk in the new lunar space suit with a built-in backpack life-support system.

Apollo 10 May 18–26, 1969. Crew: Thomas P. Stafford Jr., John W. Young Jr., and Eugene A. Cernan aboard *Charlie Brown* (CM) and *Snoopy* (LM)

Apollo 10 was a dress rehearsal for the coming lunar landing, and it accomplished all the objectives of the lunar visit except the actual touchdown. One objective was to test the LM's guidance and navigation systems in the Moon's gravitational field and its descent and ascent engines. Astronauts Stafford and Cernan brought the LM to within 9 miles (15 km) of the proposed first landing site—the Sea of Tranquility—and then fired their ascent engine. After a momentary and unexpected but controllable gyration, the LM rendezvoused and docked with the CM. Apollo 10 splashed down in the South Pacific after 192 hours (8 days) in flight. In only a matter of weeks, humans would walk on the Moon for the first time.

Gazing across the lunar terrain near Smyth's Sea, Apollo 11 crew members saw the startling view of the Earth rising above the Moon's horizon and took this photograph.

Apollo 11: "One Giant Leap"

Apollo 11 July 16–24, 1969. Crew: Neil A. Armstrong, Michael Collins, and Edwin E. "Buzz" Aldrin Jr., aboard *Columbia* (CM) and *Eagle* (LM)

A camera mounted on the mobile launch tower captured this liftoff shot of Apollo 11 at 9:32 AM (EDT) on July 16, 1969.

Apollo 11 left the launchpad on July 16, 1969. On the morning of July 20, while on the far side of the Moon, the spacecraft did not have radio contact with Houston for about 33 minutes. Although this had also happened on Apollo 8 and 10, it was an anxious time during which *Eagle,* carrying Armstrong and Aldrin, undocked from *Columbia.* As soon as radio contact was reestablished, Neil Armstrong announced, "The Eagle has wings."

Eagle flew next to *Columbia* for a final visual inspection. Then Houston gave the go-ahead for descent. Once again, the crucial maneuver took place on the far side of the Moon, out of contact with Earth. *Eagle's* descent engine fired and when it reappeared in sight, it was on its way down. Everything was "go."

Almost on the Rocks From orbit above the lunar surface, *Eagle's* descent required two engine burns and took about twelve and one-half tension-filled minutes. The first burn took the two astronauts down to an altitude of about 50,000 feet (15,240 m) above the Moon. Within five minutes, as *Eagle* descended to 6,000 feet (1,824 m), an alarm light came on. Houston said it was only a computer overload, nothing serious, and that they should continue the landing.

At about 3,000 feet (910 m) above the lunar surface, a second warning light flashed. Again Houston said to proceed with the descent. There were four more alarms within four minutes. With this distraction, the astronauts did not notice that the automatic guidance system was directing them into a field of large boulders. When they looked out the window and saw the problem, they were less than 2,000 feet (610 m) from touchdown.

This image (top) was taken from a telecast by the Apollo 11 lunar surface camera, as astronaut Neil Armstrong descended the ladder of the Lunar Module prior to taking the first step on the Moon. (The black bar running through the center of the picture is related to the television transmission.) A lunar footprint (inset) made by Armstrong's boot remains on the soft lunar soil.

At 300 feet (91 m), Armstrong took manual control and accelerated *Eagle* at about 55 miles per hour (90 km/hr) to move it to a smoother landing area. Then a warning that only 5 percent of the fuel for the descent rocket remained meant they had to either land within 94 seconds or abort the landing and return to the CM. Armstrong finally spotted a smooth site, with boulders on one side and craters on the other, but as he approached it, the lunar dust kicked up by *Eagle's* engine obscured his visibility to only a few feet.

At 33 feet (10 m) above the Moon's surface, with *Eagle* lurching dangerously, and with only about 20 seconds of descent fuel remaining, Armstrong adjusted his controls and gently landed the spider-legged LM about four miles (6 km) from the target site on the Sea of Tranquility. Armstrong's first words from the Moon were: "Houston, Tranquility Base here. The *Eagle* has landed." Armstrong's historic announcement was met with loud cheers in Houston and collective relief and exhilaration worldwide.

The Soviet orbiter Luna 15 was in orbit 11 miles (18 km) above the surface of the Moon at the time the *Eagle* landed.

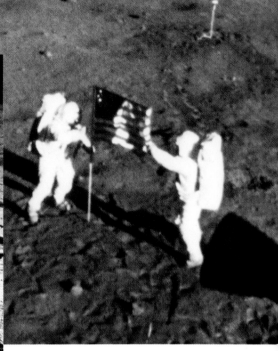

Astronaut Edwin E. Aldrin Jr. (above), Lunar Module pilot, becomes the second human to walk on the surface of the Moon. The deployment of the U.S. flag (right) on the Moon is captured for posterity during the Apollo 11 lunar landing mission.

Man on the Moon About six and one-half hours after landing, the first human touched the lunar surface. As he stepped from the LM onto the Moon, Armstrong made his historic comment, "That's one small step for [a] man, one giant leap for mankind."

Shortly after, Armstrong scooped up some lunar soil and put it in a Teflon bag in his spacesuit. If an emergency arose and *Eagle* had to take off quickly, at least that small sample would return with him to Earth.

When Armstrong had been outside the LM for almost an hour, Aldrin left the craft to join him, and the astronauts worked together for almost 90 minutes—collecting about 48.5 pounds (22 kg) of Moon rock and soil. They conducted several scientific experiments, including setting up a seismic unit to measure lunar tremors and a laser reflector that sent narrow beams of light back to Earth to measure Earth-Moon distances more accurately. Armstrong and Aldrin placed an American flag on the Moon and received a telephone call from President Richard Nixon.

After 21 hours, 36 minutes, on the Moon, Armstrong and Aldrin fired *Eagle's* ascent rocket and left what Aldrin had described as "magnificent desolation" to rendezvous with astro-

What Apollo 11 Left Behind

A plaque was left on one leg of *Eagle's* descent stage, bearing the signatures of the three astronauts and of President Richard Nixon, as well as the inscription, "Here Men from the Planet Earth First Set Foot Upon the Moon July 1969 A.D. We came in peace for all mankind." Apollo 11 also left equipment for ongoing scientific experiments; a medallion with the likeness of the Apollo 1 patch, to commemorate the ill-fated crew of Apollo 1—Gus Grissom, Roger Chaffee and Ed White; medals honoring deceased cosmonauts Yuri Gagarin and Vladimir Komarov; an inch-and-a-half (3.8 cm) silicon disc with goodwill messages from leaders of 73 nations microetched upon it; and a gold olive branch symbolizing peace.

The Apollo 11 crew is welcomed with a heavy showering of ticker tape in one of the largest parades in the history of New York City. Pictured in the lead car, from the right, are astronauts Neil A. Armstrong, Michael Collins, and Edwin E. Aldrin Jr.

naut Michael Collins in the CM. The LM's spent descent stage served as a launching pad and remained behind. The astronauts' total mission lasted for more than 195 hours. Upon their return to Earth, they were quarantined for a period of three weeks to make certain that no strange microbes had returned to Earth with them. The heroes were greeted by the president on the recovery ship and then hailed in parades and official celebrations across the nation.

Apollo 12 November 14–24, 1969. Crew: Charles "Pete" Conrad Jr., Richard F. Gordon Jr., and Alan L. Bean aboard *Yankee Clipper* (CM) and *Intrepid* (LM)

In an inauspicious beginning, Apollo 12—with the objective of precision targeting and pinpoint landing—was struck by lightning twice within the first minute of liftoff. As there was no damage, the rest of the voyage went smoothly. The landing target on the Ocean of Storms was near the site of an earlier unpiloted Surveyor 3 craft that had landed about two and one-half years before. The *Intrepid* landed about 600 feet (180 m) from Surveyor—a much less tense landing than that of Apollo 11's *Eagle*.

A flight director monitors a television transmission from Apollo 13 in the Mission Operations Control Room at the Manned Spacecraft Center, Houston, Texas. Astronaut Fred W. Haise Jr., is visible on the screen. The safe return of the three Apollo 13 astronauts (inset) was cheered all over the world.

During their 31.5 hours on the Moon—transmitting the first color TV pictures—Conrad and Bean made two Moon walks. They set up an Apollo Lunar Surface Experiment Package (ALSEP) which, on this mission, included a magnetometer to detect magnetic fields and a seismometer to measure ground movements, as well as other devices. They retrieved some pieces from the Surveyor, later found to contain Earth bacteria, which had survived two and one-half years of heat, cold, near-vacuum, and dryness. They collected a variety of Moon minerals and soil samples. On return to the CSM they sent the LM *Intrepid* crashing onto the Moon's surface at 5,000 miles per hour (8,047 km/hr). On impact the force was equivalent to an explosion of one ton of TNT and it created the first measured artificial "moonquake."

Apollo 13: Near Disaster April 11–17, 1970. Crew: James A. Lovell Jr., John L. Swigert Jr., and Fred W. Haise Jr. aboard *Odyssey* (CM) and *Aquarius* (LM)

A frighteningly troubled flight, Apollo 13 was more than 56 hours into its lunar mission and more than halfway to the Moon when, on April 13, 1970, an explosion ruptured an oxygen tank in the SM and damaged an entire panel, including all the tanks and systems inside it. This left the astronauts without oxygen, water, and power from the *Odyssey*. Fighting for their lives with help from mission control, the crew circled the Moon and headed back to Earth, using the limited oxygen and power available in the LM *Aquarius* to stay alive.

The LM was designed to sustain two people for 50 hours, but by carefully conserving both power and air, three astronauts survived in this improvised "lifeboat" for 95 hours. It was an unpleasant trip home; the LM temperature was only a few degrees above freezing and water was rationed to just six ounces per person, per day. Dehydrated and exhausted, they successfully splashed down in the Pacific on April 17—ending almost six days of extreme tension. As a NASA statement read:

> Apollo 13 must officially be classed as a failure . . . But in another sense, as a brilliant demonstration of the human capability under almost unbearable stress, it has to be the most successful failure in the annals of space flight.

Apollo 14 January 31–February 9, 1971. Crew: Alan B. Shepard Jr., Stuart A. Roosa, and Edgar D. Mitchell aboard *Kitty Hawk* (CSM) and *Antares* (LM)

After the near-disastrous Apollo 13 flight, nine months of modification and testing were done before launching Apollo 14. Alan Shepard, America's first man in space, led this mission and, at age 47, became the oldest man to walk on the Moon.

This time Shepard and Mitchell used a mechanized cart to carry their equipment on their two Moon walks. On the first, they set up another ALSEP. At one point on the second walk, Houston found that Shepard's heartbeat was too fast and irregular and ordered the first space rest period. All was well. Shepard had brought along a golf club head and two golf balls. He put the head on the end of a piece of equipment and hit one ball 200 yards (180 m) and the other about 400 yards (365 m).

The astronauts of Apollo 14 spent a total of more than nine hours gathering specimens on the lunar surface. They brought back about 96 pounds (45 kg) of lunar rock.

Apollo 14 astronaut Alan B. Shepard is shown beside the mechanized cart used on the Moon walks. For Shepard, 47 years old, inclusion on this mission was a personal triumph. He had made the first U.S. suborbital flight in 1961, then been grounded for eight years for medical reasons.

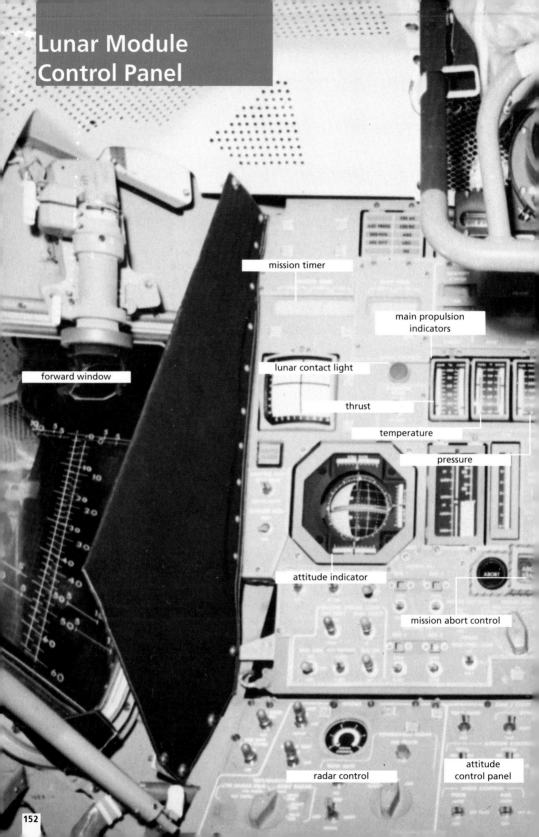

Lunar Module Control Panel

mission timer

main propulsion indicators

forward window

lunar contact light

thrust

temperature

pressure

attitude indicator

mission abort control

radar control

attitude control panel

alignment
optical telescope

forward window

master alarm

lunar contact light

On August 5, 1971, Apollo 15 astronaut Alfred M. Worden became the first human to take a walk in deep space about 197,000 miles (317,041 km) from Earth; the walk lasted about 16 minutes.

Apollo 15 July 26–August 7, 1971. Crew: David R. Scott, Alfred M. Worden, and James B. Irwin aboard *Endeavour* (CSM) and *Falcon* (LM)

With an additional battery and more life-support supplies, the redesigned Apollo LM could make longer missions. This alteration resulted in Moon exploration time being doubled to almost 67 hours.

The Rover Apollo 15 was the first mission to use the Lunar Roving Vehicle (LRV), dubbed the "rover." It was collapsible and was stowed on the side of the LM. Just pulling a cord caused it to drop to the ground and unfold. It contained navigational equipment that kept track of where the LM was located and had enough battery power to travel 55 miles (89 km), with a top speed of 7 miles per hour (11 km/hr). A rover—left behind each time—was also used on the Apollo 16 and 17 missions.

Astronaut James B. Irwin uses the Lunar Roving Vehicle during Apollo 15 lunar-surface activity. This four-wheel-drive vehicle was specially designed to ride on the lunar surface. Mount Hadley looms in the background.

The astronauts completed three EVAs, one each day for three separate days, spending more than 18 hours on the lunar surface. They traveled a total of 17 miles (27 km) in the rover, exploring the beautiful lunar canyon of Hadley Rille, plains, and deltas. They set up another ALSEP and gathered 173 pounds (79 kg) of

interesting rocks, including the "genesis" rock, thought to date from the early solar system, and a rock that contained tiny spheres of green glass.

After *Falcon's* return to *Endeavour,* the LM was crashed on the surface to calibrate all three ALSEP seismometers. Then a scientific satellite was launched from *Endeavour* into orbit around the Moon; it continued to send data to Earth for nearly a year.

Apollo 16 April 16–27, 1972. Crew: John W. Young Jr., Thomas K. Mattingly II, and Charles M. Duke Jr. aboard *Casper* (CSM) and *Orion* (LM)

Shortly after the LM undocked for the Apollo 16 Moon landing, an engine vibration developed in the CSM. When the astronauts were finally cleared to proceed, they had lost six hours but finally landed in the Descartes highlands—the highest of Apollo landing sites—18,000 feet (5,475 m) above lunar "sea level." During three EVAs Young and Duke set up another ALSEP and a small astronomical observatory to photograph ultraviolet emissions from clouds of hydrogen and other gases enveloping Earth and other celestial bodies. They also set up a cosmic ray detector to provide data about the composition and origin of the rays.

The crew used a rover to visit several craters, including the North Ray Crater—at three-quarters of a mile (1,200 m) across and 650 feet (198 m) deep, the largest crater explored by astronauts on the Moon. In over 71 hours they covered about 17 miles (27 km) and collected 213 pounds (97 kg) of lunar rocks, the largest weighing over 25 pounds (11 kg).

A Round-Trip Rock Ride

Scientists were surprised to discover, by analyzing early samples of lunar rocks brought back to Earth, that a lunar magnetic field existed. To test if the rocks had become magnetized in the spacecraft, scientists demagnetized a rock brought back by Apollo 12 and sent it back to the Moon on Apollo 16 to compare it with newly collected samples. The new rock samples proved to be magnetic, confirming a lunar magnetic field, and the test rock was left on the Moon.

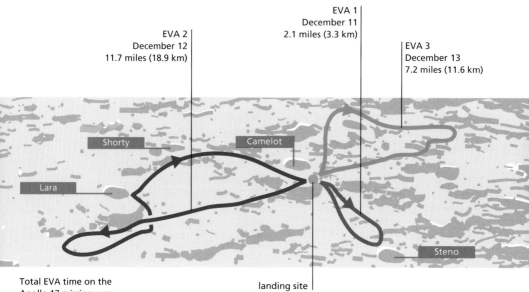

EVA 1
December 11
2.1 miles (3.3 km)

EVA 2
December 12
11.7 miles (18.9 km)

EVA 3
December 13
7.2 miles (11.6 km)

Shorty

Camelot

Lara

Steno

landing site

Total EVA time on the Apollo 17 mission was more than 22 hours, and the astronauts drove a total of 21 miles (34 km). Four of the lunar craters adjacent to the EVA paths are shown above on this computer-simulated lunar surface.

Apollo 17 December 7–19, 1972. Crew: Eugene A. Cernan, Ronald E. Evans, and Harrison H. "Jack" Schmitt (the first scientist-astronaut) aboard *America* (CSM) and *Challenger* (LM)

Challenger, with Cernan and Schmitt aboard, landed in the deep Taurus-Littrow Valley on the Sea of Serenity, just 300 feet (91 m) from the target landing site. This site was selected because recent landslides had brought material from the heights of the surrounding Taurus Mountains to within easy reach, and photographs of that area suggested evidence of relatively recent lava flows there.

A more sophisticated ALSEP was set up. It included a lunar atmospheric composition experiment and the surface electrical properties transmitter, which would help scientists understand the electrical nature of the material beneath the Moon's surface.

Cernan and Schmitt spent the longest time on the Moon—75 hours—including more than 22 hours outside *Challenger.* They collected about 244 pounds (111 kg) of lunar rocks and traveled 21 miles (34 km) in the third rover. Schmitt unexpectedly found orange soil in the crater, samples of which were found to be 3.8 billion years old. Although the soil tested volcanic, the crater in which the samples were located tested nonvolcanic.

Meanwhile, in the CSM several complex scientific instruments, though not unique to Apollo 17, were operating very successfully. Some of the best maps yet obtained of the Moon were produced. The lunar sounder provided continuous data at three different frequencies to help scientists probe the depths of the Moon, and an infrared scanner relayed information on surface temperatures and heat emission.

During the final Apollo mission, astronaut Harrison H. Schmitt demonstrates the enormous size of a split boulder on the Moon. Photographing and collecting lunar soil and rock samples were vital parts of each Apollo mission.

The Lunar Data Harvest

Study of the 842 pounds (382 kg) of rock and soil samples, the vast number of photographs, and the quantity of data sent by the ALSEPS yielded a rich harvest of information. For example, we now know that most Moon minerals are similar to Earth minerals, but many are heavier in titanium and iron. One such newly discovered mineral was named *armalcolite,* combining parts of the names of the Apollo 11 crew who returned it to Earth.

The highland lunar rock samples proved to be over 4 billion years old and are probably what remains of the Moon's original crust. The lowland rocks are over 3 billion years old, dating from the period when lava welled up from the interior and spread over the lowland plains.

The Moon was found to have a very thin atmosphere, consisting of helium, hydrogen, argon, and neon. Artificial "moonquakes" created by crashing Apollo hardware on the Moon indicated that the top lunar layer is composed of cracked and fractured rock, which explains the resulting echo or ringing. Whether or not other geological features of the Moon can be compared to Earth is still open to question. "Apollo represents a positively mythic accomplishment for the human species," said astronomer Carl Sagan. "The Moon was the metaphor of the unattainable. And, look, we had 12 people walking on the surface of the Moon. Its historic significance is really hard to overstress."

Clementine

The 500-pound (227 kg) unpiloted spacecraft *Clementine,* the first U.S. Moon mission since Apollo 17 in 1972, entered lunar orbit in February 1994. A military-civilian project, *Clementine* completed its lunar mapping mission in June 1994. *Clementine* is less than four feet (1.2 m) wide and about six feet (1.8 m) long. Its miniaturized cameras transmitted stunning, detailed images that have been pieced together to create panoramic mosaics of the Moon and even used to identify rock types.

Exploring the Planets

The Moon has been explored, but the bulk of the solar system, from mighty Jupiter to tiny Pluto, remained a mystery until robotic probes ventured to the outer planets. These unpiloted space-craft have provided exciting glimpses of the planets and their moons never seen before, but much remains for the next generation of space explorers—robotic or human—to discover.

5:07

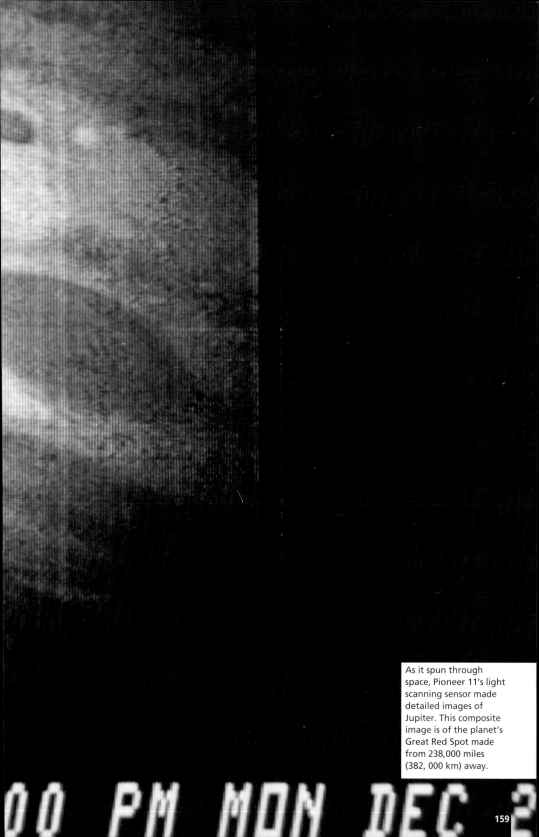

As it spun through space, Pioneer 11's light scanning sensor made detailed images of Jupiter. This composite image is of the planet's Great Red Spot made from 238,000 miles (382, 000 km) away.

00 PM MON DEC 2

Reaching the Planets

To date, humans have achieved dramatic successes in both Earth orbital and lunar missions. In exploring the planets, it has been robotic technology, however, that has accomplished amazing feats during our systematic, step-by-step exploration of the solar system.

To comprehend the time and distance of planetary missions requires a shift in our earthbound frames of reference. Distances traveled to the planets make the approximately 3-day, 240,000-mile (385,000 km) trip to the Moon seem like a brief ride. For example, when Voyager 2 encountered Neptune, twelve years into its epic journey, it had traveled nearly 3 billion miles (4.82 billion km) and was still transmitting pictures back to Earth. Furthermore, as it moves beyond our solar system, traveling at about 37,000 miles per hour (59,500 km/hr), some 320 million miles (515 million km) a year, it will take Voyager 2 about 40,000 Earth years to pass its first star!

Missions to Venus

One of the brightest objects in the heavens, Venus—named for the goddess of love and beauty—has played a role in many mythologies. Shrouded in thick, sulfurous clouds that hide its surface, Venus is difficult to explore with telescopes on Earth. But Venus has been the subject of intense exploration by spacecraft.

The size and mass of Venus, called Earth's twin planet, are virtually the same as Earth's. Also the planet closest to Earth, Venus is a natural first target for exploration. Although Venus's surface is tranquil, with slow winds and infrequent volcanism, its temperatures reach more than 850° F (470° C). Lightning illuminates the sky. Venus has an atmosphere that is 98 percent carbon dioxide, and atmospheric pressure is ninety times that of Earth. It is not a hospitable world for human exploration.

Since 1960 the former Soviet Union, with 18 robotic spacecraft, and the United States, with three, have been attempting to study Venus whenever a favorable alignment allowed for an easier, fuel-efficient trajectory to reach the planet. Learning how Earth and Venus have evolved so differently may give us vital scientific information about Earth.

Titan's thick haze layer is shown in this enhanced Voyager I image—taken on November 12, 1980—at a distance of 270,000 miles (435,000 km).

Soviet Union

Venera In an attempt to learn more about Venus, the Soviet Union achieved many "firsts" with its Venera ("Venus") program: the first Venus atmospheric probe, the first probe impact on Venus, the first analysis of the soil of Venus, and the first images from the surface of the planet. Early Soviet experience became the foundation for increasingly ambitious missions. For two decades, Soviet scientists almost monopolized Venus exploration, but they shared some of their data with the international scientific community.

Venera missions 1 through 3, launched intermittently between 1961 and 1965, were not successful space ventures. The Venera 1 probe attempted the world's first journey to another planet. Only 15 days after launch, radio contact was lost. Venera 2's flight had a similar outcome. All appeared to go well with Venera 3 until the probe entered the atmosphere of Venus and crash-landed on March 1, 1966.

Venera 4, launched in June 1967, arrived at Venus in October of that year. A capsule launched from the spacecraft into the atmosphere of Venus transmitted data regarding its chemical composition, pressure, and temperature. But, during descent, the capsule ceased functioning; its final achievement was surviving part of the descent.

Images of Venus taken by Mariner 10 are retouched, color-enhanced, and made into a mosaic by the U.S. Geological Survey. These photos were taken in ultraviolet light to demonstrate the spiral circulation patterns of Venusian clouds.

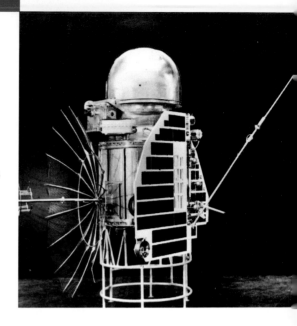

The Venera 1 Venus probe weighed 1,420 pounds (644 kg). It had a cylindrical body, and was equipped with two solar panels and a spiked radio antenna shaped like an umbrella.

Beginning with the Venera 5 and Venera 6 missions launched in January 1969, descent probes were strengthened and their descent was accelerated in an attempt to help them reach the surface of Venus before being destroyed by the enormous pressure and heat of its atmosphere. Although readings were returned from the atmosphere, both probes crashed in May and did not collect any surface data.

Launched in August 1970, Venera 7 transmitted data for 35 minutes during its descent and for 23 minutes after landing. This transmission—on December 15, 1970—was the first from the surface of another planet.

Venera 8, launched in March 1972, made a soft landing on Venus in July and did some basic surface analysis. Data, transmitted for 50 minutes, revealed that the light level on the surface of Venus was similar to that on Earth on a rainy day.

In June 1975, Venera 9 and 10 were launched. Put into orbit around Venus, these spacecraft acted as radio relay stations for the landers. In October, both probes landed, each transmitting black and white images that showed a rugged, rocky surface—the first pictures of the surface of Venus.

Venera 11 and 12 reached Venus at the end of 1978. These probes sampled the chemical composition of the atmosphere during their descent and returned data from the surface.

Venera 13 and 14 entered orbit around Venus in March 1982. The two landing craft, transmitting color images of the grey rocky surface and using drills to sample the rocks, provided the first data about the chemical composition of the planet's crust.

Venera 15 and 16 went into orbit around Venus in October 1983. Their large antennas bounced radio signals off the surface and used the returned echoes to construct radar images of the surface and to map the north polar region. Collectively, the Venera missions provided a wealth of data on the atmosphere and surface of Venus. They showed that the surface of Venus, like Earth's, has been molded by volcanic and tectonic activity. Seismic activity was also detected.

Vega Launched in 1984, the international Vega project had two phases: first, a study of the atmosphere, clouds, and surface of Venus; second, a rendezvous with Halley's Comet in 1986. Nearing Venus in June 1985, the Soviet Vega 1 and 2 spacecraft launched landing capsules. As these capsules descended they released atmospheric balloon probes, slowed by parachutes, which floated above the surface of the planet for several days collecting and transmitting data.

This project was characterized by scientific contributions from several nations including the Soviet Union, France, Germany, the United States, and various other countries. Data from both phases of the project were also tracked by a network of radio telescopes positioned around the world.

The descent capsule (left) used on both Vega 1 and Vega 2 had a coiled antenna for transmitting data. The Proton rocket (right) waits on its launch site prior to the December 21, 1984, launch of Vega 2. Vega is a contraction of the Russian words for Venus and Halley.

United States

Mariner While the Soviets pursued their planetary probes, the United States launched ten Mariner missions to explore Venus, Mars, and, last, Mercury. Mariner 1, launched toward Venus in July 1962, went off course and had to be destroyed for safety reasons. Mariner 2, a backup craft launched about a month later, became the first successful interplanetary spacecraft. Mariner 2 continued into an orbit around the Sun, sending data back to Earth for nearly a year.

Mariner 5, a modification of the Mars-bound Mariner 4, was launched in 1967. Mariner 10, the final Mariner mission, transmitted data and images of both Venus and Mercury in 1974. (See Missions to Mercury, pages 173–174.)

These Mariner missions returned atmospheric readings and provided data on the temperature and magnetic field of the planet. They confirmed that Venus has high surface temperatures. Scientists also learned that Venus has no significant magnetic field, probably due to the fact that the planet's rotation is very slow (243 Earth days).

Pioneer-Venus Two spacecraft, the last in the Pioneer series (see pages 175-177), were sent to explore Venus. Pioneer Venus Orbiter (also called Pioneer-Venus 1) was launched in May 1978. As well as recording details of the atmosphere and the solar wind relative to the planet, the orbiter mapped about 90 percent of the planet's surface. A second mission, Pioneer Venus Multiprobe (Pioneer-Venus 2), was launched in August 1978. Four probes were placed in the atmosphere on both the day and the night sides of the planet. During the probes' descent, the temperature, pressure, density, and chemical composition of the atmosphere were measured in more detail.

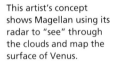

This artist's concept shows Magellan using its radar to "see" through the clouds and map the surface of Venus.

Magellan Because the surface of Venus is shrouded in clouds, studies of that planet require radar systems that can penetrate the clouds. While important radar observations had been made from Earth, it was clear that a spacecraft orbiting Venus—as Venera spacecraft had done earlier—could achieve better results.

In 1984 a U.S. spacecraft called the Venus Radar Mapper (VRM) was assembled using spare parts from the Voyager, Galileo, and Ulysses spacecraft. Renamed Magellan, after the famous explorer who led the first voyage around the world (A.D. 1519–1522), its primary objective was to map the surface of Venus.

Magellan was launched from the Space Shuttle *Atlantis* in May 1989. Fifteen months later it was in orbit around Venus, mapping the planet's surface and presenting clear evidence of volcanoes, some possibly still active. By 1992 it had met its primary objectives but continued collecting scientific data into 1994. Magellan's original goal was to map 70 percent of Venus; it actually mapped 99 percent. As a result of the Magellan mission, there is a high-resolution global map of Venus that surpasses in detail the best global maps of Earth.

Dramatic improvements in image resolution and data quality resulted from Magellan. Superseding the Venera data, the Magellan data provided highly detailed images of volcanic structures, lava plains, highland plateaus, and impact craters and ridges on the planet.

This three-dimensional radar-data image of Venus—later enlarged to reveal details—was created by Magellan's synthetic aperture radar. Magellan provided an unprecedented volume of data on Venus compiled on more than 90 CD–ROM disks.

About 100 red and violet filter images provided by Viking were used to create this mosaic of Mars. The color-enhanced image shows bright white areas to the south of the planet which are regions covered by carbon dioxide frost.

Missions to Mars

Mars—the red planet—has always been associated with the mythological god of war. It has also been thought to be the most likely planet in the solar system, other than Earth, where life in some form could have existed. In the nineteenth century, after astronomers thought they saw canals on Mars, people imagined complex, possibly Earth-threatening civilizations living there. Recent visits by unpiloted spacecraft have revealed a dry, barren, desert-like world.

Soviet Union

Mars Series The Soviets were the first to attempt to approach Mars. In late 1962, after earlier unsuccessful tries, the Mars 1 spacecraft was launched to fly past the planet. After it covered 66 million miles (106 million km), however, radio contact with Mars 1 was lost on March 21, 1963.

In 1971, Mars 2 and Mars 3 were placed in orbit around the planet. Each consisted of an orbiter and a lander which was released to touch down on the surface of Mars. A huge dust storm

The landing probe of Mars 3, launched nine days after Mars 2 in May, 1971, was mounted inside a cone-shaped shield on one end of the craft. Although the landers failed, both orbiters transmitted data about Mars until September 1972.

was in progress when the Mars 2 lander reached the surface. Within 20 seconds the lander—the first Earth object to reach Mars—lost radio contact with Earth, most likely because of the dust. Mars 3 landed a short time later but, once again, contact was lost with the probe after only 20 seconds.

In 1973 the Soviets launched four more spacecraft. Mars 4 and 5, launched in July, were orbiters; Mars 6 and 7, launched in August, carried landers. All launches were successful, but one of Mars 4's retro-rockets failed and the orbiter was lost in deep space. Mars 5 entered Mars orbit and operated successfully for 20 orbits, transmitting data and images during 10 of those orbits. The Mars 6 lander crashed, and a trajectory error caused the Mars 7 lander to miss the planet.

Mars 5 was the only successful spacecraft of the four, and it confirmed earlier measurements made by Mars 2 and 3. The Soviets suspended their Mars program for fifteen years and concentrated on Venus instead. But, in a sign of international cooperation, the images from Soviet Mars 5 and U.S. Mariner 9 were combined to produce a comprehensive atlas of Mars.

Phobos In the 1980s, when Soviet scientists had reached the limits of their technology at Venus, they renewed interest in Mars with the Phobos project. The objective of this program was to land two probes on the asteroid-like Martian moon Phobos to examine it. This mission also had international support, carrying scientific experiments from several other countries. Two identical spacecraft, Phobos 1 and 2, were launched in July 1988.

Each Phobos spacecraft carried two small landers and instruments for observing both the Sun and Phobos. The first mission was intended to anchor itself on Phobos, then act as a "permanent" geophysical station. The second, known as the Hopper, would study the chemical composition of Phobos's surface in different areas by landing, taking measurements, then "hopping" to another location. The Hopper would also generate vibrations so the lander could measure the density of Phobos's surface and other geophysical characteristics.

Phobos 1 never reached Mars. Phobos 2 entered Mars orbit early in 1989 and began to transmit pictures. But, as preparations were made to send the spacecraft toward Phobos and to release the two lander capsules, contact was lost. The mission failures were later attributed to software error and computer malfunctions.

United States

Mariner Launched in November 1964, Mariner 4 was the first successful mission to Mars, flying past the planet in July 1965. It recorded 22 images of the planet, revealing a moonlike, crater-covered surface. In addition, instrument readings showed the atmosphere of Mars to be very thin, compared to Earth's, and composed mostly of carbon dioxide. Atmospheric pressure at the surface measured less than one percent of Earth's.

Mariner 6 and 7 were launched in 1969 on flyby missions. Together they gathered a total of 201 images of the planet. Mariner 6 took measurements of the structure and composition of the atmosphere, images of the surface, and measurements of surface temperature. Mariner 7's high-resolution images revealed a polar ice cap consisting of dry ice, haze, and clouds; it also found evidence of wind weathering and water erosion.

After a launch failure of Mariner 8, Mariner 9 was launched in May 1971, becoming the first spacecraft to orbit another planet. Initially a dense dust storm on Mars prevented the cameras from seeing anything. The orbiter, however, tracked the

This replica of the Phobos probe was presented at an International Air Show at Le Bourget, France. Each Phobos weighed 6.8 tons (6.2 MT); they were launched from the Baikonur cosmodrome.

storm, making images of it on each orbit. When the dust began to clear some months later, Mariner 9's two television cameras transmitted more than 7,300 images, covering about 90 percent of the Martian surface. Mariner 9 also viewed the two small Martian moons, Deimos and Phobos.

Mariner 9 spent almost a year in orbit, using complex radar and thermal measurements to map large areas of the planet. It showed that Mars has two very distinct hemispheres: a southern hemisphere, with a very old cratered surface, and younger surface features in the northern hemisphere. Other features, such as volcanic structures surpassing in size any on Earth, were also found. A huge canyon, dry river channels, lava flows, and polar regions consisting of ice and frozen carbon dioxide (dry ice) suggest that Mars had a warmer, wetter period early in its history.

But Mariner 9 could not answer the ultimate question of whether life existed on Mars. The only way to find out for sure was to land there—the objective of two Viking spacecraft.

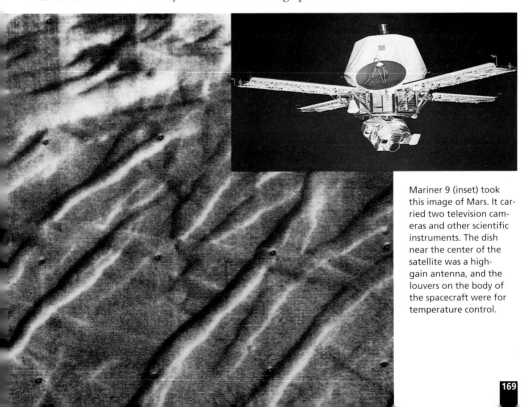

Mariner 9 (inset) took this image of Mars. It carried two television cameras and other scientific instruments. The dish near the center of the satellite was a high-gain antenna, and the louvers on the body of the spacecraft were for temperature control.

Viking The two Viking spacecraft, each consisting of an orbiter carrying a lander, were highly sophisticated probes launched in 1975. The Viking orbiter, based on the Mariner 9, was enlarged to carry a lander. Because it took nearly 20 minutes for Earth/Mars radio signals to reach the spacecraft, the landers' descent to the surface of Mars was controlled by onboard computers. The orbiters relayed data from the landers to Earth.

As soon as Viking 1 entered orbit around Mars in June 1976, it began sending pictures used by mission controllers to select touchdown sites for the landers. Viking 1 made the first successful soft landing of a robot spacecraft on Mars on July 20, 1976. Within minutes, its cameras were activated and images of the surface were received on Earth, showing a rust-colored landscape of rocks and boulders with a reddish sky overhead. Viking 2 returned a similar view from the other side of the planet.

By the summer of 1980, the Viking orbiters had returned more than 3 years of weather reports and imaged almost 100 percent of the surface of Mars. The gravitational field and atmospheric water vapor were studied, and a thermal map of the surface was made. Surface temperatures ranged from −20° F (−29° C) in the afternoon to −120° F (−84° C) at night. During the Martian winter, a layer of frost formed on the ground. Both landers used their robotic arms to collect samples of Martian soil, which were analyzed in the lander. Chemical analysis at these sites could not give evidence of past or present life elsewhere on Mars—a question still unanswered.

The Viking missions confirmed that Mars is red because the soil contains oxidized iron and the surface is more arid than any Earth desert, with water restricted to a possible layer of per-

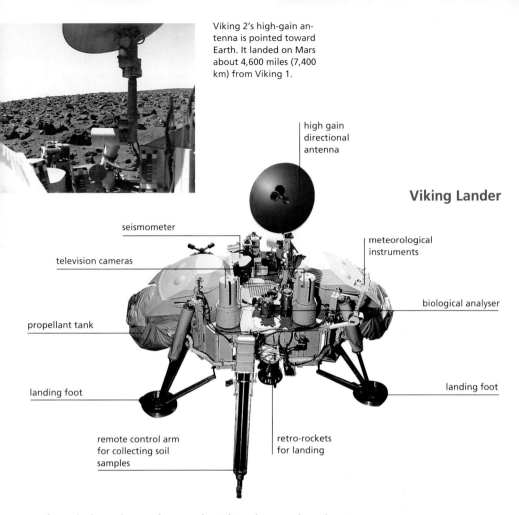

Viking 2's high-gain antenna is pointed toward Earth. It landed on Mars about 4,600 miles (7,400 km) from Viking 1.

Viking Lander

high gain directional antenna

seismometer

meteorological instruments

television cameras

biological analyser

propellant tank

landing foot

landing foot

remote control arm for collecting soil samples

retro-rockets for landing

mafrost below the surface and within the north polar ice cap. The Martian atmosphere contains elements necessary for sustaining life, such as nitrogen, oxygen, carbon, and argon. However, the Martian atmosphere is principally carbon dioxide and unsuitable for human life.

The first spacecraft to conduct prolonged research on the surface of another planet, the Vikings were very productive. The Viking orbiters operated until mid-1980, and the landers operated until 1980 and 1982, all far exceeding their planned lifetimes. More than 4,500 images were transmitted by the landers; over 52,000 images were transmitted by the orbiters, in color and stereo.

**"Well, we are the Martians now."
—science-fiction writer Ray Bradbury after Viking 1 had touched down on Mars.**

Mars Observer After 15 years of examining the wealth of data from Viking missions, NASA launched the Mars Observer from Cape Canaveral on September 25, 1992—sending a robot spacecraft toward a polar orbit around Mars to complete studies of its weather, climate, and surface. Such thorough exploration has intrinsic scientific value and also is a prerequisite for possible piloted missions.

The mission was planned for the duration of one Mars year (687 Earth days), with a scheduled arrival at Mars in August, 1993. As the spacecraft prepared to enter Mars orbit after its eleven-month journey, contact was lost and attempts to reestablish tracking and communications signals were unsuccessful. An engineering investigation concluded that the most probable cause of failure was a propellant system rupture during the orbit maneuver that left the spacecraft disabled or destroyed.

Mariner 10

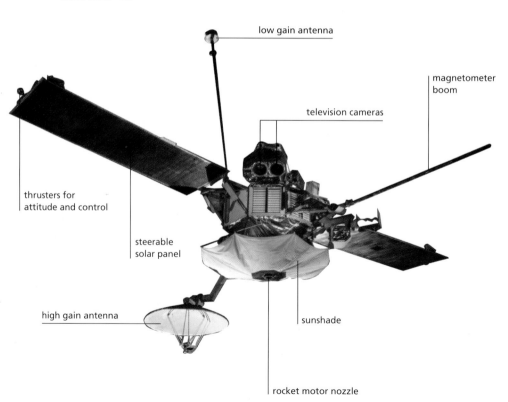

low gain antenna

magnetometer boom

television cameras

thrusters for attitude and control

steerable solar panel

high gain antenna

sunshade

rocket motor nozzle

Missions to Mercury

Mercury is a small planet in our solar system—about one-third the size of Earth. Of the planets, only Pluto is smaller. The closest planet to the Sun, Mercury is difficult to observe from Earth because of the Sun's glare. Before the Mariner 10 mission in 1974, only basic facts about Mercury's size, mass, and orbital variables were known from telescopic observation and mathematics. We now know its period of rotation—that is, one complete day on Mercury—is almost 59 Earth days. It has the fastest orbital speed in the solar system, orbiting the Sun once every 88 days.

Mariner 10 Mariner 10's objective was to explore Venus, then Mercury, in a single journey. Once every ten years the relative positions of Venus and Mercury are such that a space probe launched from Earth toward Venus can be propelled in a slingshot orbit toward Mercury by Venus's gravity. Using gravity this way saves both time and fuel and allows the spacecraft to carry a heavier payload. An opportunity for such a mission occurred in 1973.

Although similar in design to Mariner 6 and Mariner 7 spacecraft, Mariner 10 was adapted to resist the high temperatures close to the Sun—at midday on Mercury, with the Sun overhead, the temperature can reach 800° F (420° C). The main body of the spacecraft was protected from the Sun's heat by a shade, thermal wrapping, and rotating solar panels designed to maintain the temperature at about 239° F (115° C).

Two identical television cameras and other instruments were mounted on a steerable platform, fixed to the main body of the spacecraft, or carried on retractable booms. A rocket motor made trajectory corrections possible.

Mariner 10 was launched in November 1973. It eventually completed four flybys—one of Venus and three of Mercury. It

The Mariner 10 spacecraft obtained this view of Mercury during its outbound pass on March 19, 1974. This photo mosaic has been tinted to approximate the visual appearance of the planet.

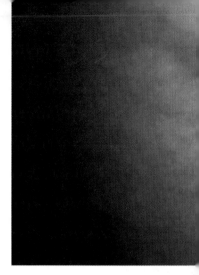

Mariner 10's camera was powerful enough to take a legible photo of classified ads from a quarter of a mile (400 m) away.

arrived at Venus in February 1974 and sent some 3,000 images to Earth, including many in ultraviolet light, which gave insight into the circulation of the clouds. Data were also gathered on the planet's atmosphere and magnetic field.

From Venus, Mariner 10 continued on a seven-week voyage to Mercury, reaching that planet in March 1974. During its trip past Mercury, it took images of about 40 percent of the planet's surface. Those photos showed the surface of Mercury to be heavily cratered, much like the surface of the Moon.

Mariner 10 provided many facts about Mercury. The chemical composition of Mercury's negligible atmosphere is sodium, potassium, helium, and hydrogen. Mercury has a large iron-rich core and extreme temperatures ranging from 800° F (420° C) to –300° F (–180° C). Mariner 10 also provided new data on the properties of Mercury's surface and its possible magnetic field.

After the flyby of Mercury, Mariner continued to orbit the Sun, circling it in 176 days, or twice the length of Mercury's year. This brought the spacecraft back near Mercury for the second look in September.

On its third flyby in March 1975, Mariner passed close enough to Mercury to collect valuable data on the planet's newly discovered magnetic field. Eight days later the fuel used to control the spacecraft's flight ran out and contact with the probe terminated.

The Mariner mission was successful in several ways. First, many images and measurements were made. More important, for the first time, a space probe visited two planets, using the gravity of one to propel it to the other. This slingshot technique was used later with the Voyager probes to the outer planets. Also, Mariner 10 provided new information on the atmospheric pressure, surface temperature, and magnetic field of Mercury. To date, no other missions to Mercury are planned.

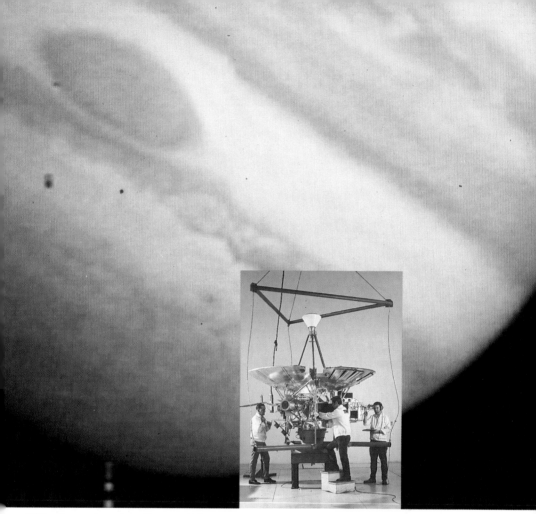

Missions to the Outer Planets

Pioneer Pioneer 10 and Pioneer 11 both traveled into interplanetary space, becoming the first spacecraft to pass beyond the orbit of Mars. They were designed to examine the asteroid belt and sample the environment of Jupiter and, for Pioneer 11, Saturn. The robotic spacecraft were identical; each was equipped with a system that would monitor its own instrumentation, transmit data to Earth, and receive modifying commands from Earth.

Launched in March 1972, Pioneer 10 became the first spacecraft to fly beyond Mars. Scientists weren't certain that a spacecraft could safely cross the asteroid belt. This belt is an area that extends 270 million miles (435 million km) along the flight path between Mars and Jupiter and contains asteroids ranging from the size of gravel to several hundred miles in diameter. But, 135 days after launch, Pioneer 10 completed its passage through the belt without incident. The closest the spacecraft passed to a known asteroid was 5.5 million miles (8.8 million km).

Technicians make final adjustments to the Pioneer 10 spacecraft (inset) before launching. Pioneer 10 transmitted the first detailed images of Jupiter's cloud tops—taken from a distance of 1,305,000 miles (2,100,000 km).

Pioneer 10's Plaque

Pioneer 10 carried a plaque with information about Earth for any intelligent life it might reach. The drawing shows a man and woman, the man's hand raised in a sign of peace. Behind them is the outline of the spacecraft. Along the bottom is a representation of the solar system, showing the location from which the spacecraft came. On the left are lines representing the positions of 14 pulsars and the center of the galaxy in relation to the Sun.

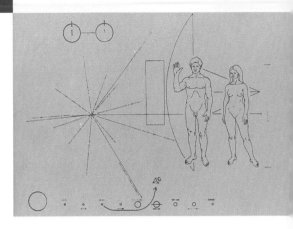

In December 1973 the spacecraft encountered Jupiter. Because the giant planet is a half-billion miles (805 million km) from Earth, the task of transmitting pictures and data to Earth was a significant technical challenge requiring a much longer than usual communications chain between the spacecraft and the ground computers. NASA relied on the Deep Space Network—which includes the 210-foot (64 m) antenna at Goldstone, in California's Mojave Desert, as well as similar antennas located near Madrid, Spain, and Canberra, Australia.

Outside the craft, instruments were extended on a 21-foot (6.4 m) boom to measure the interplanetary magnetic field originating from the Sun. One instrument facing the Sun measured the energy, direction, and number of solar wind particles. Two particle telescopes examined nuclei moving through space and identified eight possible elements ranging from hydrogen to oxygen.

Instruments onboard Pioneer 10 took the first close-up images of Jupiter and its moons—about 500 images whose quality far exceeded any previously taken from Earth. But even more important was the new information Pioneer discovered about Jupiter. Data revealed that its magnetic field is 2,000 times stronger than Earth's. Pioneer 10 also confirmed that Jupiter has no detectable solid surface.

After leaving the vicinity of Jupiter, Pioneer 10 crossed the orbit of Pluto and, on June 13, 1983, it became the first spacecraft to begin exiting the solar system. In the event of future discovery, Pioneer 10 carries a plaque with information about Earth and its inhabitants attached to its antenna mount.

Data are still being received from Pioneer 10, despite the fact that it is now on its way out of the solar system, close to three

billion miles (4.82 billion km) from Earth. Radio signals from the probe take nearly 14 hours to reach mission control. It is expected that communications from Pioneer 10 will continue until about the year 2000.

Pioneer 11, launched in April 1973, also crossed the asteroid belt without sustaining any damage and went on to Jupiter. The spacecraft flew close enough to use Jupiter's gravity to change its course for Saturn. Pioneer 11 transmitted images of Saturn's rings, even discovering two new rings. Pioneer 11 also revealed significant new knowledge about Saturn's magnetic field, indicating that it has a core about the same size as Earth's.

Voyager The Voyager spacecraft for NASA missions to Jupiter and the other outer planets were much more complex than those of the earlier Mariner and Pioneer missions. Because of the distance they would travel from Earth, they had to be sophisticated enough to detect errors and make corrections before Earth-based controllers even knew about them. Due to the distance they would be from the Sun, nuclear power generators replaced solar panels. Each Voyager weighed 1,820 pounds (826 kg)—more than three times that of the Pioneer spacecraft—and would use the gravitational force of one planet to propel it to another. Voyager probes transmitted data to Earth at a rate 100 times faster than the Pioneer 10 and 11 probes.

On February 13, 1979, Voyager 1 took this photo of Jupiter showing the Great Red Spot and the planet's moons Io (left) and Europa. On the Voyager spacecraft (inset), the cameras and other instruments that must be aimed are located on the science boom at the right. The twelve-foot antenna in the center focuses the communication between the spacecraft and controllers on Earth.

The Grand Tour

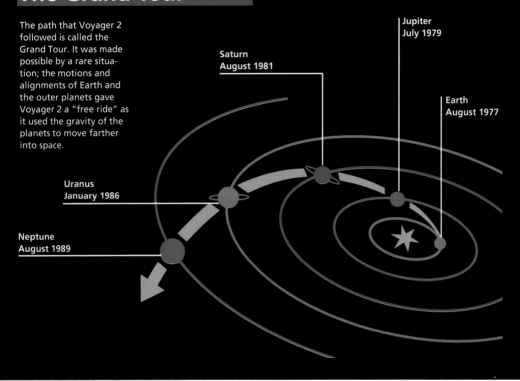

The path that Voyager 2 followed is called the Grand Tour. It was made possible by a rare situation; the motions and alignments of Earth and the outer planets gave Voyager 2 a "free ride" as it used the gravity of the planets to move farther into space.

Jupiter
July 1979

Saturn
August 1981

Earth
August 1977

Uranus
January 1986

Neptune
August 1989

Voyager 2 was launched first, on August 20, 1977, followed by Voyager 1 sixteen days later. Voyager 1 took a more direct route and arrived at Jupiter in March 1979, four months before Voyager 2. It immediately began transmitting detailed images of the planet, as well as data on its magnetic field. It discovered a thin ring around Jupiter and imaged the planet's major moons in detail for the first time. Passing within 216,800 miles (348,900 km) of Jupiter, Voyager 1 allowed a close examination of the planet's atmosphere and the characteristics of the Great Red Spot—a complex, permanent storm rotating in a counter-clockwise direction.

Both spacecraft used Jupiter's gravity to place themselves on a trajectory toward Saturn. Voyager 1 reached Saturn in November 1980 and, once again, transmitted spectacular images and data. Pictures of Saturn's rings showed that each major ring consisted of thousands of ringlets made up of countless ice and dust particles. Images revealed new details of the known moons of Saturn and featured several new ones, bringing the total to more than 20. Voyager 1 also rendezvoused with Saturn's largest moon, Titan—revealing that Titan has a thick, smoglike atmosphere that con-

The Voyagers' Music and Messages

As each Voyager spacecraft journeys ever farther from Earth in interstellar space, it carries a gold-plated-copper phonograph record encoded with images of Earth, greetings in about 60 languages, and more than an hour of music compiled from cultures all over our planet. One of the greetings is from President Jimmy Carter, who says in part, "We hope someday to join a community of galactic civilizations. This record represents our hope and our determination, and our good will in a vast and awesome universe."

The "Sounds of Earth" recording is mounted on Voyager 2, then shielded in aluminum for protection during the flight of the spacefcraft past Jupiter and Saturn and out of the solar system.

ceals its surface and a surface temperature of –300° F (–184° C). Then, passing Saturn, Voyager 1 headed out of the solar system.

These successes led to a decision to extend the mission of Voyager 2 to Uranus and Neptune. Nine months later, in August 1981, Voyager 2 flew past Saturn and used its gravity to go on toward Uranus. In January 1986 Voyager 2 made its closest approach to Uranus, returning many images of the planet, its faint ring system, and its five largest moons. It also measured the length of a day on Uranus—16 hours 48 minutes. Voyager 2 revealed 10 new moons of Uranus, bringing the total to 15, before departing for Neptune, where it arrived three and one-half years later. The spacecraft gathered data on Neptune's ring system and its largest moon, Triton, and discovered six more new moons.

Voyager 1 embarked on the last phase of its mission: transmitting images of the view looking back through the solar system. Our knowledge of the universe has been enriched by discovery of moons that were unseen by telescopes on Earth and are now added to the "population" of the solar system. We have enjoyed a closer look at our wondrous solar system—yet found no other planet quite like Earth.

Galileo After falling victim to many delays and setbacks, Galileo was launched in 1989 for a scheduled arrival at Jupiter in 1995. Originally called the Jupiter Orbiter Probe, Galileo has two components: an orbiter and an atmospheric probe. The orbiter will explore the radiation belts of Jupiter, as well as its moons. It will then orbit the planet after releasing the probe into Jupiter's atmosphere. The probe will measure the atmosphere's composition and perform other scientific measurements as it descends through Jupiter's thick cloud layers.

Galileo (inset) is being prepared in the Vertical Processing Facility at the Kennedy Space Center to be joined with the Inertial Upper Stage booster. Galileo (below) is launched from the Space Shuttle *Atlantis* on October 18, 1989.

Flight planners developed a six-year flight plan that required that Galileo be launched against, not in, the direction of Earth's motion around the Sun. As Galileo fell toward the Sun, it was speeded up by gravitational forces from a close flyby of Venus. Then it headed back to Earth, again using gravitational forces to make a long elliptical pass into the asteroid belt, and do another Earth flyby before heading toward Jupiter.

Before Galileo reaches Jupiter, its two components will separate. Just before the probe enters the atmosphere, the orbiter will fly by Jupiter's moon Io to make observations and use Io's gravity to slow its approach and help it into the proper orbit around Jupiter. As the probe slams into Jupiter's atmosphere at about 112,000 miles per hour (180,200 km/hr), a heat shield will protect its instruments from the heat of the atmospheric friction and a parachute will slow its descent. Information on the atmosphere, cloud layers, lightning, and energy absorption will be collected and transmitted to the orbiter overhead, which will then relay it to Earth.

Extremes of pressure and temperature eventually will destroy the probe. After relaying to Earth the last information transmitted by the probe, the orbiter will continue to circle the planet and transmit data to Earth for nearly two years. Galileo's orbiter will encounter the four biggest and brightest of Jupiter's 16 known moons, with special attention to the composition of their surfaces.

Galileo has already begun to provide scientists with new data on the comet Shoemaker-Levy 9, large fragments of which hit Jupiter at great force in July 1994.

The Galileo spacecraft (above) is seen just after deployment. It is named for Galileo—the seventeenth-century Italian astronomer and physicist. Galileo was the first person known to use a telescope to study the planets and stars.

Comets

Comets whose orbits bring them close to the Sun make particularly interesting targets for space exploration. The most famous comet is Halley's Comet, which reappears every 76 years. When Halley was due to pass near the Sun in 1985 to 1986, the international scientific community became involved in an unprecedented cooperative effort to explore it. Soviet, Japanese, and European probes were sent to intercept Halley, and other scientific spacecraft and telescopes were used for observation. The International Halley Watch, involving about 800 scientists from 40 countries, coordinated worldwide observations.

Vega: Phase 2 Leaving Venus using the slingshot effect of its gravity, Vega 1 and Vega 2 continued on their planned journey to rendezvous with Halley. On March 6 and 9, 1986, they became the first probes to fly past Halley at an encounter distance of 4,800–5,400 miles (8,000–9,000 km). Vega 1 transmitted 500 images and detailed scientific data on the structure and chemistry of the comet. When Vega 2 arrived three days later, it collected more images and data. The valuable information from these two probes was used by the European Space Agency (ESA) to make last-minute course adjustments to its Giotto probe.

Giotto The Giotto probe, ESA's first interplanetary mission, encountered Halley a week after Vega. This mission was the joint effort of 87 institutions and about 200 scientists.

Giotto approached within 376 miles (605 km) of Halley's core, closer than any other probe. Giotto determined that the icy, rocky core of the comet was 7.5 miles (12 km) long and shaped somewhat like a peanut. Giotto carried various instruments including a camera to record images of the comet's nucleus and inner coma, or atmosphere.

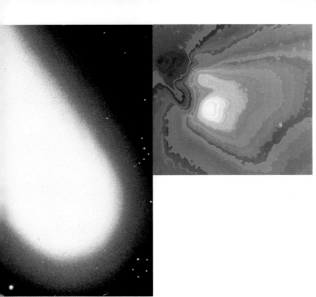

This image was made when Giotto encountered Halley's Comet on March 13, 1986—at a distance of 12,430 miles (20,000 km) from the nucleus. The different color levels correspond to different light levels from white to black. The nucleus is in the upper left corner.

Sakigake and Suisei The first two Japanese interplanetary spacecraft, Sakigake and Suisei, were launched in January and August 1985, to study Halley. Because there was concern that material in the comet might impact and damage the spacraft, neither approached the comet as closely as Vega or Giotto. However, they did gather useful information on the solar wind around the comet.

ISEE 3 (ICE) In 1978, the International Sun-Earth Explorer was launched. Intended to study conditions in space close to Earth, ISEE 3 was the first spacecraft to orbit around a particular point in space (called a *halo orbit*) rather than around a celestial body. The point was located between Earth and the Sun, approximately one million miles from Earth.

From its point between Earth and the Sun, the satellite was later moved into a heliocentric orbit designed to pass the comet Giacobini-Zinner in September 1985. ISEE 3 then was programmed to encounter Halley in March 1986 and to compare the electromagnetic fields of these two comets.

Since the satellite had not been originally designed to study comets, it relayed no images. However, it was able to transmit data on the interaction between the comets and the solar wind. ISEE 3 was the first spacecraft to fly through a comet's tail. For this phase of its mission, the ISEE was renamed the International Cometary Explorer (ICE).

Using Space

Ancient astronomers studied the same skies as twentieth-century stargazers—until the 1957 launch of Sputnik, the first artificial satellite, turned those skies into a new frontier. Now, hundreds of orbiting satellites facilitate global communications, making the world a smaller place. Will wider cooperation between peaceful nations make functioning "space colonies" a reality in the next century? Only time will tell.

Astronauts aboard
the Space Shuttle
Discovery prepare to
secure the retrieved
Palapa B-2 satellite on
November 12, 1984.

The Echo II telecommunications satellite (inset) is shown undergoing inflation stress tests before being launched into orbit on January 25, 1964. More sophisticated communications satellites were in service by August 1984, when Syncom IV, shown here against a backdrop of the African coastline, was released from the Space Shuttle *Discovery*.

Satellite Systems

Hundreds of satellites designed for a variety of commercial, scientific, and military purposes are circling Earth. A representative sample of satellite missions launched by the United States and other nations is described in this chapter.

Telecommunications

Communication satellites are responsible for instantaneous television coverage of events all over the globe, for the speed of international banking and finance, for the almost-immediate delivery of vast amounts of information, and for international e-mail and telephone service to all parts of Earth. Almost any transfer of information that depends on cables, lines, or antennas can now be communicated via satellite. One large communications satellite can carry—simultaneously—over 100,000 phone calls and several television signals.

How They Work Because of Earth's curvature, high frequency (HF) radio systems require aerials placed every 31 to 62 miles (50–100 km) . But one satellite relay station can link Africa and Asia or Europe and North America. A communications satellite is equipped with receivers, amplifiers, and transmitters which enable it to send and receive thousands of transmissions at the same time. If it circles Earth in a geosynchronous orbit—completing one circuit every twenty-four hours—it will stay over one point on Earth's surface. A network of such satellites thus can provide global coverage.

Comsat and Intelsat As the value and commercial potential of satellite communications became apparent in the early 1960s, the United States created the Communications Satellite Corporation (Comsat) in 1962. Comsat's purpose was to establish a global communications satellite system. This led to the formation in 1964 of Intelsat (the International Telecommunications Satellite Organization) by eleven countries. It had a three-part mission: to establish a global communications satellite system owned jointly by many nations, to pursue technological advances, and to become a permanent organization. Since then Intelsat has grown to include over 130 nations that share ownership and operation of six generations of Intelsat stationary satellites. This constellation of Intelsat satellites over the Atlantic, Pacific, and Indian oceans relays telegraph and TV transmissions and over two-thirds of all international phone calls.

Telstar 401, the first of three AT&T advanced communications satellites is readied for shipment to Cape Canaveral. It was launched on December 15, 1993, on an Atlas IIAS rocket. Telstar 401 entered commercial service February 1, 1994.

Echo, Telstar, and Relay In October 1945, science fiction writer Arthur C. Clarke first suggested a space-based telecommunications device in the journal *Wireless World*. Ten years later John Pierce, of Bell Telephone Laboratories, approached NASA with his ideas about telecommunications satellites. In 1960 Echo, an aluminized plastic balloon 100 feet in diameter, was launched into a low Earth orbit where it functioned until 1968 as the world's first passive reflector (bouncing rather than actively transmitting voice, picture, and radio signals back to Earth) in communications experiments. Echo II was launched in 1964 and operated until 1969.

AT&T developed Telstar, the world's first commercial satellite; its first launch was in 1962. Telstar was an active-repeater (transmitting or retransmitting) satellite. At the same time, NASA was developing another active-repeater satellite called Relay. The two were intended to demonstrate the feasibility of low-orbit, multi-satellite communications systems for telephone, TV, and data transmissions. A Telstar was launched as recently as 1993.

Syncom Syncom is a geostationary satellite developed by Hughes Aircraft and NASA, first launched in 1963 and 1964. Syncom weighed half as much as either Telstar or Relay and had an extra internal rocket stage so it could be launched into a twenty-four-hour orbit inclined to the equator. Syncom's success made it the model for communication satellites. Since 1984, several Syncoms have been launched from the Space Shuttle.

Astronauts on the STS-49 *Endeavour* mission successfully capture the Intelsat VI telecommunications satellite on May 13, 1992. Then the satellite was attached to a booster that lifted it to its corrected, higher orbit.

Molniya The Soviet Union needed a system of satellites and ground stations that could provide TV, telephone, and telegraph links over its sprawling northern frontiers. Because a satellite in orbit above the equator has difficulty communicating with places nearer the polar regions, the Soviets needed a different orbital track for their satellites. Since the first launch of Molniya ("lightning") in 1965, three generations of Molniya have been the answer. Several Molniya satellites are placed in a twelve-hour orbit. Essentially each is turned on in sequence as it approaches its apogee and turned off as it approaches its perigee.

In 1971, the Soviets and their Communist bloc partners formed a communications consortium called Intersputnik, in response to Intelsat.

Palapa During the 1970s and 1980s other regional satellite systems developed. For Indonesia, Palapa A (no longer in use) and B were launched to connect isolated Earth stations in southeast Asian islands and countries. The Palapa C program will cover India, Japan, and China.

Arabsat Among the Arab League nations, Arabsat was developed to provide both regional and international connections. Like their European neighbors, Arabsat members are also plugged into Intelsat. The first two Arabsat satellites were launched in 1985; a second generation is planned for launch in the 1990s.

The Doppler Effect

If you listen to a passing siren, it will sound higher as it approaches and lower as it moves away. During the siren's approach, each successive sound wave is emitted closer to you, so you hear more sound waves per second (a higher frequency). As the siren moves away, you hear fewer sound waves per second in a given time (lower frequency). This phenomenon, called the Doppler effect or Doppler shift, affects not only sound waves but any kind of wave, including visible light and radio waves. Because of this, Doppler shifts can be used to measure the velocity of objects moving toward or away from an observer.

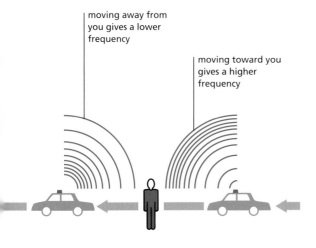

moving away from you gives a lower frequency

moving toward you gives a higher frequency

Marisat Marisat satellites were developed for communications between ships and shore stations. During the first few years after Marisat's launch in 1976, the U.S. Navy was its primary user. By 1980, however, there were more than 300 commercial users, including oil tankers and cargo ships. In 1982 control of the Marisats was transferred to Inmarsat—the International Maritime Satellite Organization. It now handles search-and-rescue messages, navigational information, medical assistance, and routine ship-to-shore communications.

Westar Westar is a series of U.S. telecommunications satellites operated by the Western Union Telegraph Company. They are designed to provide continuous video, data, and voice communications throughout the United States, Puerto Rico, and the Virgin Islands. Since 1974, six Westars have been launched, including one that was retrieved by the Space Shuttle when it failed to achieve its proper orbit.

Oscar OSCAR is the designation given to an orbiting satellite that carries amateur radio. Since 1961 there have been many OSCARs including several launched in the 1990s. In 1965 OSCAR 3 became the first free-access communication satellite, and amateurs in over sixteen countries communicated through it. Amateur satellite operators sponsor themselves through AMSAT (the Radio Amateur Satellite Corporation, a nonprofit group).

The cesium or rubidium atomic clocks within Navstar/GPS satellites are so accurate that they will lose no more than one second every 300,000 years.

By the end of March 1991, a total of 15 Navstar/GPS satellites like this one were in service over the Persian Gulf region to provide maximum coverage during the military conflict there.

Navigation

In 1957, several scientists measured Sputnik's radio signal frequency and discovered that by measuring the Doppler effect they could plot the satellite's orbit. Knowing the exact orbit of a satellite and the Doppler shift of its radio signal, it is possible to calculate a satellite's exact position on Earth.

Transit Faced with the need for an improved navigational system for nuclear submarines, scientists obtained funding and began work in 1958 on the Transit program, which used the Doppler effect to locate positions on Earth. In 1967, Transit entered the commercial market where it has since been used to determine positions for offshore oil exploration, fishing, surveying, and navigating.

Navstar/GPS Spurred to improve the accuracy of Transit, the U.S. Navy and Air Force began work on Navstar/GPS (Navigation Satellite for Time and Ranging/Global Positioning System). This system uses twenty-four navigational satellites in three 12,500-mile (20,100 km) orbits at 55 degrees inclination. The satellites complete one orbit every twelve hours. Large receivers are used aboard ships and airplanes, and handheld receivers have also been developed. These receivers can identify the navigator's position within 50 feet (15 m) in three directions and can estimate the speed at which a craft is traveling within a fraction of a mile per hour.

COSPAS-SARSAT Used in ground and sea search-and-rescue missions, COSPAS-SARSAT is an international satellite system established in 1979 as a joint venture among Canada, the United States, France, and the Soviet Union. Four polar satellites receive distress signals and transmit them to local user terminals around the world. These terminals send location data to a mission control center, and rescue authorities are notified. The system, capable of determining a position to within 1.2 miles (2 km), has saved thousands of lives.

Meteorology

Satellites have been invaluable aids to meteorologists. They provide views of weather patterns, indicate temperatures through infrared measurements, and scan immense areas of Earth.

One important tool for gathering meteorological data from space is the *radiometer*. A radiometer measures the intensity of radiation and is the key instrument used to map relative temperatures on Earth's surface.

TIROS The first significant meteorological satellite was TIROS (Television and InfraRed Observation Satellite). First launched in April 1960, it recorded developing cloud formations, storm systems, and cloud cover images. The 19 satellites in the first two generations of the TIROS program were launched between 1960 and 1969. Their images were invaluable in weather forecasting, especially in the advance warning of violent storms.

Current TIROS satellites, such as the Advanced TIROS–N (above), collect global weather data continuously. For a large percentage of people on Earth, though they may not realize it, TIROS data are essential to their daily lives.

NOAA Starting in 1970 the fourth-generation TIROS, the NOAA (National Oceanographic and Atmospheric Administration) series, was launched. NOAAs are polar-orbiting satellites that have remained operational well beyond their expected two-year life cycle. They can view almost all of Earth's surface twice a day. NOAAs have provided weather images and data concerning atmospheric humidity, snow, ice cover, and temperature of the atmosphere and sea surface. They also relay COSPAS-SARSAT distress signals.

SMS In 1974 and 1975 the United States launched SMS 1 and 2 into geostationary orbit over the equator to provide ongoing near-hemispheric coverage. SMS (Synchronous Meteorological Satellite) specialized in obtaining data from remote receivers such as ocean buoys, river gauges, ships, balloons, and aircraft.

GOES, Meteosat, Insat, and GMS In 1977 the U.S. GOES, European Meteosat, and Japanese GMS satellites were launched into geostationary orbits. Their task was to observe cloud formations and measure movement in the atmosphere. Data from these satellites are used to determine large-scale mass density and wind distribution.

In 1987, these satellite series were coordinated into a network. They maintain a geostationary orbit at an altitude of about 22,300 miles (36,000 km) along the equator. GOES East and West (GOES West quickly failed) cover North and South America, the Pacific, and the Caribbean; Meteosat covers Europe and Africa; India's Insat covers the Indian Ocean and parts of Asia; and GMS covers Southeast Asia, Australia, and the western Pacific region.

In 1994 GOES 8 was launched carrying upgraded instrumentation for measuring heat and humidity and for imaging weather patterns. Other GOES launches are planned.

An Atlas I rocket that holds the GOES 8 weather satellite is prepared for launch. The geostationary satellite, successfully launched from Cape Canaveral in April 1994, was developed by NASA for NOAA.

Remote Sensing

Remote sensing is a way to acquire data from a distance, with no direct contact. Astronomers, ecologists, geologists, meteorologists, and the military use remote sensing devices and techniques. Remote-sensing satellites provide data on the planets, weather, and missile sites, for example. Many remote sensor detectors now used for civilian and scientific purposes are adapted from military reconnaissance technology.

Earth Resources Monitoring

Landsat The major U.S. remote land-sensing satellite program is called Landsat. Designed for topographic and geological exploration, Landsat was born out of a 1966 request to NASA from the Department of the Interior to develop a satellite to help in assessing and managing natural resources. The launch in 1972 of ERTS (Earth Resources Technology Satellite) 1—soon renamed Landsat 1—was followed by several successful launches. Landsats 4 and 5 continue to operate. In the late 1990s they are scheduled to be replaced with Landsat 7.

Currently two Landsats circle Earth every 99 minutes, making about 15 orbits every day in a near-polar orbit at an altitude of 438 miles (705 km).For data consistency, the orbit is synchronized with the position of the Sun to allow for the same Sun angle at the same latitude every day. Landsat instruments can see in both visible and infrared light with a resolution of about 36 yards (33 m) over a path about 115 miles (185 km) wide.

Landsat has been highly versatile: it has provided forecasts of global wheat production, identified Mayan archeological sites in the Yucatan, revealed fault lines that indicate oil and mineral deposits, and identified sites for dams and nuclear power plants.

SPOT The European Space Agency Earth observation program, called SPOT (Satellite Pour l'Observation de la Terre), was developed by France with support from Sweden and Belgium. First launched in 1986 by an Ariane rocket, the three SPOTS have been very successful. They operate at about 517 miles (832 km) altitude and provide complete images of the Earth every 26 days. The image area covered is approximately 38 miles (61 km) on each side.

This Landsat 4 photo (bottom left) taken in September 1982 shows the Gulf Coast in southern Louisiana and Mississippi. The bluish white crescent is the city of New Orleans. A Landsat satellite (below) is prepared for launch.

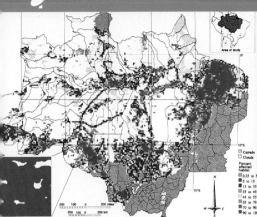

This image, of the Brazilian Amazon Basin was produced from data from the Landsat 4 and 5 satellites. It shows areas damaged by deforestation over a period of one year.

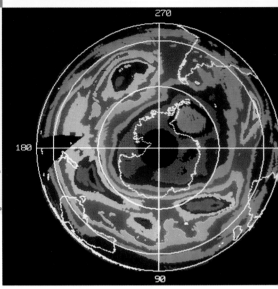

The Total Ozone Mapping Spectrometer (TOMS) provided data from August through October 1986 that was used to produce this image of the Antarctic Ozone Hole.

ERBS Another remote sensing satellite is the Earth Radiation Budget Satellite (ERBS) developed by NASA to measure radiation from both the Sun and Earth. ERBS was launched by the Space Shuttle in 1984. Its instrumentation measures ozone and other elements in the atmosphere, crucial data for understanding global climate change.

EOS EOS (the Earth Observing System) is an international remote sensing system still in its planning stage. Its aim is to study and monitor global environmental change. Late in the 1990s, coordinated launches are planned by NASA, Europe, and Japan to put observation platforms into Sun-synchronous polar orbits at approximately 435 miles (700 km) altitude.

TOMS-Meteor 3 TOMS (Total Ozone Mapping Spectrometer) is designed to measure the ozone layer of Earth's atmosphere. The first TOMS instrument was launched in 1978. TOMS-Meteor 3, a Soviet Meteor-3 meteorological satellite carrying a U.S. TOMS, was launched in August 1991 from the Plesetsk launch site by a Soviet Cyclone rocket. Its launch date put the satellite in position to observe the yearly formation of the Antarctic ozone "hole," which typically occurs between late August and mid-November. TOMS-Meteor 3 provides data on atmospheric ozone, aerosols, and radiative transfer.

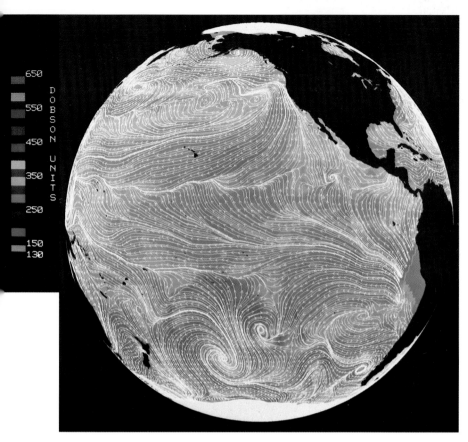

650
550
450
350
250
150
130

D
O
B
S
O
N

U
N
I
T
S

NASA's Seasat satellite provided the 1978 data used to create this image showing global wind speed (as colors) and direction (as arrows). Wind speeds on this image range from blue to gray to red to yellow.

Oceanography

Seasat Seasat was launched in June 1978 but operated for only about 100 days due to a power failure. It measured wave height, sea-surface temperature, wind direction, and wind speed using microwave, as well as visual and infrared, sensors for all-weather capability (unhindered by cloud cover). Seasat data provided a base for the TOPEX mission.

TOPEX/Poseidon The U.S. TOPEX (the Ocean Topography Experiment) was combined with France's Poseidon project, and the TOPEX/Poseidon satellite was launched in August 1992. Its mission is to develop maps of ocean topography with nearly perfect accuracy. Its microwave detectors, for example, allow the craft to "see" and measure ice floes during the long polar nights. Its radar enables it to measure wind and waves.

This mission is being carried out in cooperation with the World Ocean Circulation Experiment. In order to create a three-dimensional ocean current model, traditional observations will be combined with satellite data.

Geophysics

Satellites that provide information about Earth's physical processes and phenomena contribute to the study of geophysics.

Lageos A small, dense, spherical satellite launched by NASA in May 1976, Lageos is covered with 426 laser reflectors that are used to measure distances from the satellite to Earth stations or between points on Earth with extreme accuracy. Lageos II was launched from the Space Shuttle in October 1992.

OGO and ISEE Six OGOs (Orbiting Geophysical Observatories) were launched by NASA from 1964 to 1969 to study geophysical and solar phenomena in Earth's magnetic field and atmosphere. In 1977, twin International Sun-Earth Explorers began a detailed exploration of Earth's magnetic field.

Astronomy and Astrophysics

Astronomy in space provides better visibility without the interference of Earth's atmosphere, moisture, and clouds. From space it is also possible to study the full range of the spectrum, including the types of radiation that are blocked or absorbed by the atmosphere of Earth.

Hipparcos Hipparcos, ESA's High Precision Parallax Collecting Satellite, was launched in August 1989 to observe and measure the exact location of 120,000 selected stars. Its mission lasted for two-and-one-half-years and the results were used as a basis for additional investigations.

The Hubble Space Telescope The Hubble Space Telescope was the first of NASA's four planned Great Observatories to be launched. Named for Edwin Hubble, the astronomer who confirmed that the universe is expanding, it was dubbed the "new

The Hubble Harvest

Data from the Hubble Space Telescope include closer looks at active galactic nuclei, a galaxy with a starry pinwheel and 42 spherical star clusters, a black hole in galaxy M84, and star evolution in globular clusters. In June 1994, the telescope relayed images from the constellation Orion of stars probably in the process of forming planetary systems. The Hubble image at right is of the core of galaxy M100, tens of millions of light years from Earth.

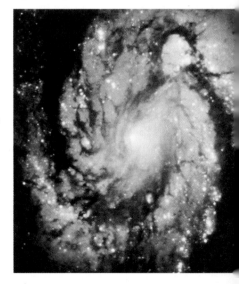

window on the universe." It makes observations in both visible and ultraviolet light of very faint and distant celestial objects, provides improved distance measurements, and investigates the evolution of stars and galaxies.

Launched by the Space Shuttle *Discovery* in April 1990, the Hubble was soon discovered to have a serious flaw; the 96-inch (2.4 m) primary mirror was improperly curved. Although the flaw was less than a hair's width, it was enough to blur the images. In December 1993, the crew of the Space Shuttle *Endeavour* serviced the Hubble by replacing solar arrays, gyroscopes, and other equipment and by installing corrective optics and a new camera. Images received in January 1994 showed that the repairs were a complete success.

Astronaut Jeffrey A. Hoffman is anchored on the end of the Remote Manipulator System arm while on an EVA during the Hubble Space Telescope servicing and repair mission. NASA plans to keep the telescope working with periodic in-space maintenance.

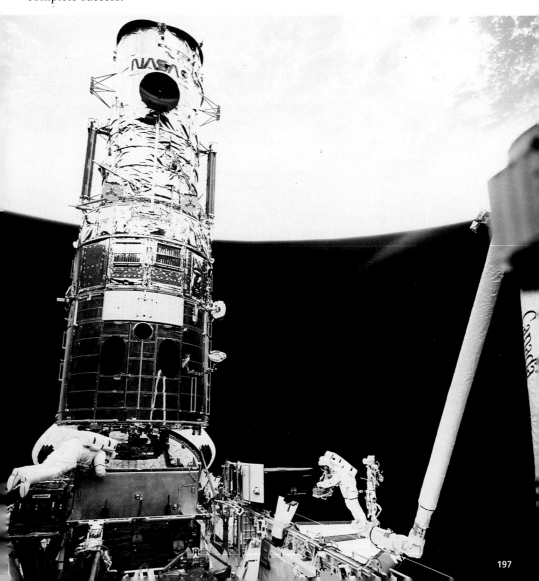

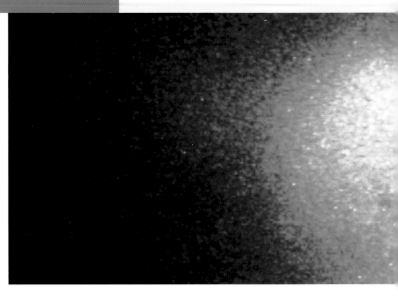

Solar Physics

The best place to observe the Sun is above Earth's atmosphere because the atmosphere absorbs or blocks much of the Sun's energy. Instruments used for solar observation as well as astronomy include *photometers*, which measure light intensity, and *spectrometers*, which measure spectral wavelengths.

This 1967 ultraviolet picture transmitted in digital form by NASA's Orbiting Solar Observatory IV shows the Sun as it had never been seen before. It reveals the Sun's atmosphere 100,000 miles above the surface.

OSO The Orbiting Solar Observatories (OSO) were the first satellites used to conduct a systematic study of the Sun. Their purpose included measuring frequency and energy of solar radiation and studying the solar atmosphere and solar flares. From 1962 to 1975 NASA developed eight satellites for the program. Initially these satellites were aimed directly at the Sun, but the increased capabilities of OSO-4 through OSO-8 allowed them to scan and make excellent images of the outer parts of the solar atmosphere.

Dynamics Explorer In early August 1981, two Dynamics Explorer satellites, DE-1 and DE-2, were launched. They investigated the interaction of energy, electric currents, electric fields, and plasmas between Earth's magnetosphere, ionosphere, and upper atmosphere with special studies of the auroras. DE-1 operated until November 1990; DE-2 reentered Earth's atmosphere in February 1983.

Hintori and Yohkoh Two Japanese satellites have observed solar X rays. Launched in 1981, Hintori ("firebird") collected data on the eleven-year cycle of solar activity. The Yohkoh ("sunshine") space probe was launched in 1992 carrying X-ray imagers and X-ray and gamma-ray spectrometers. Yohkoh has supplied new information on the Sun's corona and solar flares.

This image of the Crab Nebula (left) was made by NASA's High Energy Astronomy Observatory 2 (HEAO–2). It shows a pulsar, or pulsating star, in the center. The first HEAO (above) is being prepared for launch in 1977.

X-Ray Astronomy

X-ray astronomy specializes in observing high-energy radiation with telescopes above Earth's atmosphere. Initial observations in the early 1960s indicated that X-ray astronomy could be the key to understanding supernovas, quasars, and black holes.

HEAO HEAO-1 was launched in 1977 for an X-ray survey and mapping mission. It added to the catalog of X-ray sources begun about seven years before by Uhuru, the first X-ray astronomy satellite. The Einstein observatory HEAO-2 (High Energy Astronomy Observatory) was one of these important missions to investigate the universe in X-rays and gamma rays. Launched in November 1978 and operational until 1982, it discovered thousands of X-ray sources, including quasars believed to be more than 14 billion light years away.

Plans for the next generation of X-ray observatories include the Advanced X-ray Astrophysics Facility (AXAF). About one hundred times more sensitive than the HEAO, the AXAF will attempt to create a comprehensive X-ray map of the whole sky, observe X-rays in the Sun's corona, and chart unknown quasars.

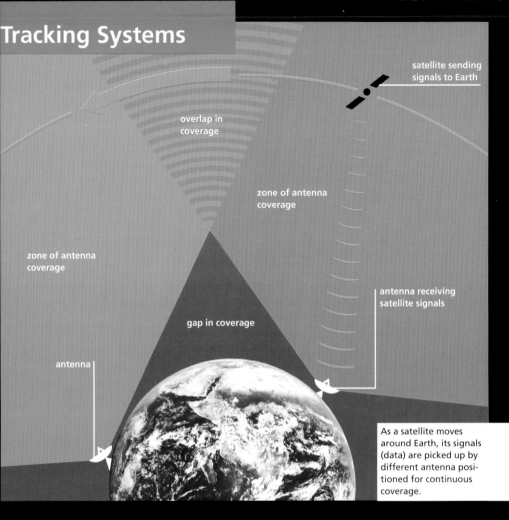

satellite sending signals to Earth

overlap in coverage

zone of antenna coverage

zone of antenna coverage

antenna receiving satellite signals

gap in coverage

antenna

As a satellite moves around Earth, its signals (data) are picked up by different antenna positioned for continuous coverage.

Once a satellite achieves a successful orbit, it is tracked by a series of ground stations that receive the satellite's signals, pinpoint its location, and calculate maneuvers needed for orbital corrections. NASA, the European Space Agency (ESA), and the French agency CNES each maintain a tracking network. Operating cooperatively, they form the Tracking, Telemetry and Command (TT&C) network of tracking stations that, along with other networks, provides international coverage of satellites in orbit.

The network uses a variety of antennas. The largest dish antennas, which can be as much as 230 feet (70 m) across, are de-signed to pick up the lowest frequencies, useful in detecting the weak transmissions from satellites, and also from space probes traveling into the distant reaches of the solar system. The antennas of the network are distributed around the globe in such a way as to minimize "blank" periods, when a satellite is out of range.

The ground stations of the TT&C pass information to control facilities. For NASA this control facility is located at the Goddard Space Flight Center, in Maryland. The control facilities store vast amounts of data and carry out sophisticated image processing based on information provided by satellites and space probes.

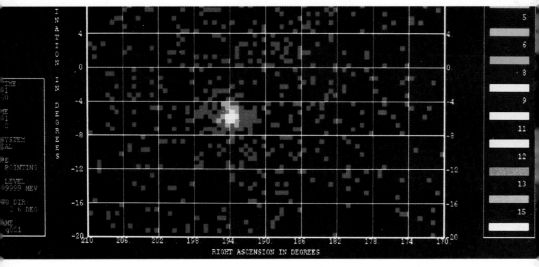

The following text appears within the image (instrument readout labels on the left side):

INATION IN DEGREES

RIGHT ASCENSION IN DEGREES

Gamma-Ray Astronomy

Gamma rays comprise the most energetic radiation in the electromagnetic spectrum. They are emitted from nuclear processes; in astronomy they are the signature of exploding stars, quasars, and pulsars.

Cos B ESA's Cos B satellite was launched in August 1975 to explore the structure, emissions, and sources of the most energetic of gamma rays—those greater than 30 million electron volts. Although Cos B had a life expectancy of only two years, it operated for almost seven.

Compton Gamma Ray Observatory In April 1991 the Space Shuttle *Atlantis* deployed the Compton Gamma Ray Observatory (GRO) to investigate celestial sources of gamma rays. Part of NASA's Great Observatories program, the Compton Gamma Ray Observatory has delivered new data on supernovas, young star clusters, pulsars, possible black holes, and quasars. It has also detected gamma-ray bursts of unknown origin.

The Energetic Gamma Ray Experiment Telescope (EGRET) on NASA's Compton Gamma Ray Observatory made this digitized gamma-ray sky map. The bright spot in the image is the quasar 3C 279 which is located in the constellation Virgo, some six billion light years from Earth.

Nanotechnology

Nanotechnology, the technology of very small systems, probably will be a focus for research, growth, and investment in the next century. Smaller, lighter satellites requiring smaller launchers and less fuel could greatly reduce launch costs. Miniature, orange-sized satellites could possibly replace huge ones.

Ultraviolet Astronomy

IUE Scientists have known for some time that hot stars, super-novas, and quasars all emit most of their energy in the ultraviolet range of the electromagnetic spectrum and that the Sun, Jupiter, Saturn, and Saturn's moon Titan are also sources of extreme ultraviolet light. But astronomers were unable to examine these sources until 1978, when the IUE (International Ultraviolet Explorer) satellite was launched. Images from the successful IUE continue to add to those of the Hubble Space Telescope.

EUVE The next generation of ultraviolet telescopes was launched in June 1992 on the EUVE (the Extreme Ultraviolet Explorer). Its instruments include four telescopes and one spectrometer. Its mission is to survey the whole sky to locate sources of extreme ultraviolet radiation.

Infrared Astronomy

IRAS Infrared radiation is characteristic of cool matter in space—for example, dust, planets, and "dark matter"—that is not hot enough to radiate much visible light. Infrared-emitting disks of dust observed around stars are thought to be sites where planets may be forming, and clouds of dust and gas in galaxies to be sites of new star formation. In 1983, the IRAS (Infrared Astronomy Satellite) was launched to detect infrared radiation from cosmic sources. A joint project of the United States, Britain, and the Netherlands, IRAS made the first infrared all-sky survey and discovered more than 200,000 sources, mostly cool stars. Among its most important discoveries were a class of infrared ultra-luminous galaxies and infrared disks around many stars.

This image of the Andromeda Galaxy was produced by combining observations made by IRAS in three infrared wavelengths. Blue represents the warmest, green is intermediate-temperature, and red is the coldest material. The telescope of IRAS had to be cooled to absolute zero to suppress its own infrared emission. It operated for almost a year until its coolant evaporated.

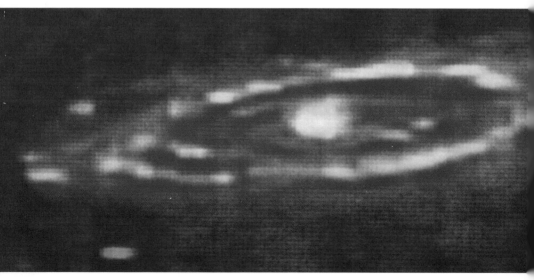

Cosmology

Cosmology is the study of the origin, evolution, and structure of the universe. The Big Bang theory is the focus of much cosmology research today.

COBE The COBE (Cosmic Background Explorer) was launched by NASA in 1989 to examine the faint microwave background radiation observed uniformly in every direction—the afterglow of the origin of the universe, according to the Big Bang theory. COBE, which operated until 1994, measured the spectrum of the background radiation and its uniformity and determined that its characteristics agree with theoretical predictions. Thus COBE provided observational confirmation of an important element in the Big Bang theory.

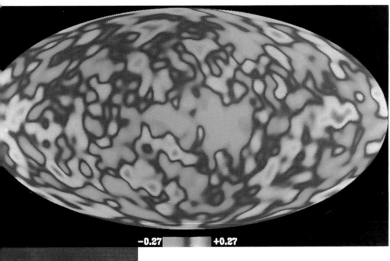

−0.27   +0.27

This microwave map of the whole sky was made from one year of data provided by COBE. Computer analysis of the data shows that the pattern of temperature variations supports the Big Bang theory.

The Big Bang Theory

The explanation of the origin of the universe favored by many scientists is the Big Bang theory. According to this model, all matter and radiation in the universe originated in an explosion at least 15 billion years ago. Stars, galaxies, planets, and other celestial objects evolved as the universe rapidly expanded and gradually cooled. This theory accounts for the observed expansion of the universe, the microwave background radiation, and the abundance of helium in the universe.

This SDI map makes Integrated Tactical Warning and Attack Assessments—that is, all components are designed to work together to assess whether warning or retaliation is required. Most SDI findings are classified and cannot be broken down for further analysis.

Military Satellites

In August 1957, Nikita Khrushchev announced that the Soviet Union had ballistic missiles capable of crossing the ocean. Even though the United Nations had proposed outer-space peace treaties, the major powers had military objectives in space. Most of the technology used in space programs came from military research, including the rockets and detectors used for scientific and commercial satellites.

Although the original impetus for military activity in space (US/Soviet competition) has waned, military organizations maintain that space-based assets are vital to any post–Cold War strategy. The new Space Applications and Warfare Center built by the U.S. Air Force in Colorado Springs, Colorado, represents considerable commitment. Clearly space still presents an important stage for military endeavors.

Early Warning and Detection

Vela Hotel Beginning in October 1963 the United States launched the Vela ("watchman") satellites to detect any violations of the 1963 nuclear test ban treaty—nuclear tests conducted in the atmosphere or in space. The first six of the Vela satellites, which were launched in pairs, were equipped with external X-ray detectors and internal neutron and gamma-ray detectors. The third pair also had an optical nuclear-flash detector. Velas 7 through 12 also monitored solar activity, providing radiation warnings for piloted missions including the Gemini extravehicular activities (EVAs) and the Apollo lunar missions. They also observed terrestrial lightning and celestial radiation. The last six Velas operated for more than ten years, ending in September 1984.

The SDI Shuttle Pallet Satellite (SPAS-II) is released by *Discovery's* Remote Manipulator System (RMS) to perform data collection on the April-May, 1991 mission. This scene, with the blue and white of Earth in the background, was taken from inside *Discovery's* crew cabin. An SDI concept—Brilliant Pebbles, or BP, space-based interceptors (inset)—were never put into service.

SDI/BMDO The Strategic Defense Initiative (SDI), nicknamed "Star Wars," was authorized in March 1983 by President Ronald Reagan. SDI was planned to be a multi-tiered U.S. military defense system using both space-based and ground-based defenses to thwart an attack from intercontinental ballistic missiles. In May 1993, SDI Organization (SDIO) was officially reorganized and renamed the Ballistic Missile Defense Organization (BMDO). Its major goal is to develop new ground-based radar and interceptor systems. A space-based part of BMDO, the Brilliant Eyes (BE) system, has been considered for deployment after the turn of the century. If developed, BE satellites would track vehicles launched into space and provide targeting data for ground-based interceptors.

Military Communications

Long-distance communications satellites for the military were introduced in the late 1960s by the United States and the Soviet Union and in the 1970s by NATO and the United Kingdom. Capabilities of these satellites have greatly improved. For example, the U.S. Navy's FltSatCom (fleet satellite communication) system provides for world-wide high-priority communications between naval aircraft, ships, submarines, and ground stations.

Navigation

The purpose of military navigational satellites is to determine the position of ground troops, aircraft, missiles, ships, and submarines accurately. Navigation has been greatly advanced by Navstar and GPS, discussed earlier in this chapter.

Nuclear Detection

Satellites are used to enforce nuclear test bans by monitoring the skies for nuclear blasts. Any intense flash could be an indication of a nuclear explosion and be the cause for additional surveillance. The Integrated Operational Nuclear Detections System is a second-generation guardian aiding the Vela satellites. These detectors are orbited on Navstar satellites.

Electronic Intelligence and Spy Satellites

Since the end of the Cold War, much information that had been classified about electronic intelligence satellites has become available. Nicknamed "ferret" satellites, they are used to listen in on secret military radio and radar transmissions. The first U.S series, called Rhyolite, has been followed by increasingly effective spy satellites, with their own code names. Usually, spy satellites are launched into a low orbit where they maneuver close to Earth's surface.

Discoverer/Corona In 1958 President Eisenhower assigned the CIA the task of developing a film recovery reconnaissance satellite program. Existing Discoverer satellites were used as "cover" for the Corona spy operation. On August 18, 1960, a capsule containing reconnaissance photos of the Soviet Union taken by Discoverer 14 was ejected from the satellite and successfully retrieved midair by a piloted aircraft. The capsule contents were then sent to Washington, D.C. This began a new era of photographic spying for the United States. Discoverer/Corona missions ended around 1962 after providing images of the Soviet Union, which were used to estimate existing Soviet ICBMs, for example.

Keyhole Instead of dropping its exposed film for retrieval, the Keyhole (KH) satellites developed the film onboard and then scanned and transmitted images to Earth. From 1963, Keyhole camera technology has evolved and been improved to the point that infrared, radar, and electronic intelligence can be transmitted to the ground almost instantaneously, as was done during the 1990–91 Persian Gulf conflict. Although the exact-name designations are unclear, it appears that the aging KH-11 is one of the reconnaissance satellites still in use.

New-generation spy satellites have sensors capable of picking up and recording electronic communications between anyone, almost anywhere, on Earth.

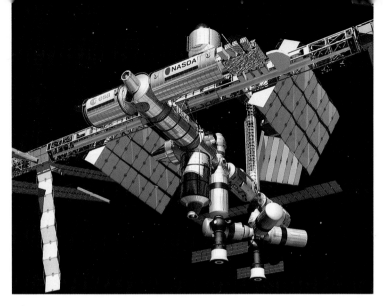

An artist's representation illustrates future international cooperation in space. Phase III of the projected International Space Station is depicted with elements provided by NASA and the Russian space agency.

Possible Future Developments

An International Space Station

The space station formerly called Freedom, then redesignated Alpha, has been planned for completion early in the twenty-first century. Initially under development by the United States, with components provided by Canada, Japan, and members of the European Space Agency, the space station gained a new partner in 1993 when Russia agreed to participate in the project. The United States has negotiated for time on the Russian space station Mir to use it as a laboratory in preparation for building a new space station.

Assembly of the space station will be preceded by a series of flights by U.S. astronauts and Russian cosmonauts on board the U.S. Space Shuttle and the Mir space station. After the construction of the core of the space station, modules provided by other participating countries may be added.

The space station could be used to carry out scientific experiments, especially biomedical research related to long-duration spaceflight, and to develop state-of-the-art technologies. It could also function as a solar power station, beaming energy to Earth. It could provide a way to maintain and repair experiments and spacecraft, or, it could function as an intermediate station for launching spacecraft farther into space.

Moon Bases and Lunar Colonies

The brief Apollo stays of one to three days inspired visions of travel to the Moon for longer periods—months or years—and residence there in either small bases or large colonies. Some would-be lunar colonists envision prospecting for and mining

This futuristic concept of a lunar outpost includes a construction shack for crew use. Other concepts include underground modules to provide protection from harsh solar and cosmic radiation. Robots could construct some parts of the lunar colonies.

lunar resources—hydrogen, helium, minerals—for commercial or industrial uses on Earth or in space. Others would use the Moon as a launchpad for missions into deep space, as it takes less energy to launch large spacecraft from the Moon's weaker gravity than from Earth. Astronomers think that the Moon would be an ideal site for telescopes and observatories. Future Moon colonists would need to draw upon lunar resources, as well as technology brought from Earth, to establish life support and power systems and an artificial environment to live in.

On to Mars?

Since the 1960s, Mars mission studies have examined various methods for human exploration of this neighboring world. Transporting people and supplies over the 40 million miles (64 million km) would not be easy. Robotic precursors would likely scout the territory and pave the way for humans. Because of the magnitude of effort and expense involved, the most feasible way for people to get to Mars may be by an international, cooperative mission.

Dreams or Reality

Whether Moon bases and Mars colonies, space cities, and voyages to other planets remain in the realm of dreams and science fiction or become reality depends on many variables. Undoubtedly, clear goals and sustained political and economic support will be needed to meet the challenges of more ambitious endeavors in space. As the successful Apollo program proved, technological hurdles can be overcome if human spirit, and political will, are committed to the effort. With such commitment who can imagine what remarkable adventures in space may define future generations?

A4 The original name of the V2 rocket. The A4 was preceded by the A1 and A2, experimental rockets developed in Germany in the 1930s by Wernher von Braun and associated engineers. *See* **V2.**

accelerometer An instrument, as in a spacecraft's navigational system, for sensing changes in speed.

Advanced X-Ray Astrophysics Facility (AXAF) *See* **Great Observatories.**

Agena U.S. rocket upper stage that was successfully combined with first-stage Thor, Atlas, and Titan rockets to launch space probes and satellites. During the Gemini VIII mission an Agena was used as the first space-docking target.

airlock An airtight chamber that separates areas of different air pressure or environment on a spacecraft and through which crew members enter or leave, once the pressure or environment in the chamber has been equalized with the area toward which they are headed.

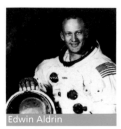
Edwin Aldrin

Aldrin, Edwin E. "Buzz," Jr. (b. 1930) U.S. astronaut; set a space walk du-

ration record for the Gemini project, in November 1966. Aldrin was lunar module pilot on the Apollo 11 mission and the second person to walk on the Moon.

America The name of the Apollo 17 command module.

Anders, William A. (b. 1933) U.S. astronaut; pilot of the Apollo 8 mission.

Antares The name of the Apollo 14 lunar module.

apogee The point in the orbit of a satellite that is farthest from Earth. *See* **perigee.**

Apollo program A lunar exploration program developed in answer to the challenge of President John F. Kennedy, who, in a speech in May 1961, called for the United States to land an astronaut on the Moon by the end of the 1960s. Drawing on the knowledge and experience gained from the Mercury and Gemini projects, NASA undertook a series of spaceflights that accomplished the program's goal with the Apollo 11 mission of July 1969. The program continued until 1972. Some of the significant missions were the following:

• **Apollo 1** (January 27, 1967) This first mission came to a tragic end on the launchpad when the capsule caught fire during a countdown test, killing astronauts Grissom, White, and Chaffee.

• **Apollo 8** (December 21–27, 1968) Astronauts Borman, Lovell, and Anders became the first humans to orbit the Moon, and the first to view the Moon's far side. They transmitted television images of the lunar surface back to Earth.

Apollo 11

• **Apollo 11** (July 16–24, 1969) A historic mission conducted by astronauts Armstrong, Collins, and Aldrin. On July 20 Armstrong and Aldrin, descending in the lunar module *Eagle,* landed on the Sea of Tranquility and became the first humans to walk on the Moon. They returned to Earth bearing the first lunar rocks and soil.

• **Apollo 12** (November 14–24, 1969) The second lunar landing, on the Ocean of Storms, during which astronauts Conrad, Gordon, and Bean performed major scientific experiments on the Moon and collected lunar samples.

• **Apollo 13** (April 11– 17, 1970) On the way to a third lunar landing, an explosion in the service module forced astronauts Lovell, Swigert, and Haise to make an emergency return to Earth, using the lunar module *Aquarius* as a lifeboat.

Apollo 13

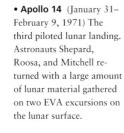

Apollo 15

- **Apollo 14** (January 31–February 9, 1971) The third piloted lunar landing. Astronauts Shepard, Roosa, and Mitchell returned with a large amount of lunar material gathered on two EVA excursions on the lunar surface.

- **Apollo 15** (July 26–August 7, 1971) This mission, with astronauts Scott, Worden, and Irwin, was the first to use the lunar rover and the first to explore lunar rills and mountains, in the Hadley-Apennine region.

- **Apollo 16** (April 16– 27, 1972) Astronauts Young, Mattingly, and Duke spent 20 hours and 14 minutes exploring the lunar surface.

Apollo 17

- **Apollo 17** (December 7–19, 1972) The sixth and final piloted lunar landing. Cernan and Schmitt explored the Moon for 22 hours before rejoining Evans in the command module and returning to Earth. It was the longest Apollo mission (301 hours) and the longest time spent on the Moon (75 hours).

Apollo-Soyuz The first joint mission between the U.S. and the USSR, completed in July 1975, in which an Apollo and a Soyuz spacecraft successfully docked together in space. This mission was the final flight of an Apollo capsule and the final launch of a Saturn rocket.

Aquarius The name of the Apollo 13 lunar module.

Ariane A family of three-stage rockets (Ariane I–V) first developed by the European Space Agency to launch its own commercial communications and weather satellites, and now providing launch services to customers from other countries. The first Ariane was launched in December 1979.

Neil Armstrong

Armstrong, Neil A. (b. 1930) U.S. astronaut; test pilot in 1960 for NASA's X-15 rocket plane and commander of Gemini VIII and Apollo 11 missions. Armstrong was the first person to set foot on the Moon, marking the occasion with his now-famous comment "That's one small step for man, one giant leap for mankind," which was televised live to the millions watching on Earth.

artificial satellites Any of various scientific or technological objects placed in orbit around Earth or another celestial body. They are used to observe, photograph, and obtain data about Earth and other planets, the Sun, and other objects in space or the environment around them. Military intelligence, navigation, meteorological forecasts, remote sensing, and communications—all depend heavily upon artificial satellites.

ascent engine A rocket on the ascent stage of a lunar module that is fired to separate that stage from the descent stage and return it to the command module.

ascent stage The upper section of a lunar module. It separates from the descent stage and returns the crew from the lunar surface to lunar orbit for rendezvous with the command module.

asteroid Any of the rocky celestial bodies, measuring from a few yards to a few hundred miles in diameter, that orbit chiefly in the asteroid belt. Some asteroids periodically cross Earth's path in their orbit around the Sun.

asteroid belt The interplanetary region between Mars and Jupiter in which most asteroids are located.

astronaut A person with the highly specialized training required to fly, navigate, or travel and work aboard a spacecraft.

astronautics The technological and scientific study of spaceflight.

Atlantis The fourth Space Shuttle. Its first flight was on October 3, 1985.

Atlas A single-stage rocket with one main engine, two booster engines, and two vernier engines, all powered by liquid oxygen and kerosene. Atlas rockets have been combined with upper-stage Agenas and Centaurs to launch space probes, military payloads, large communications satellites, and weather satellites.

atmosphere A gaseous envelope surrounding Earth and many other celestial bodies. Gravity holds an atmosphere close to the surface of the body it surrounds. Flight beyond the atmosphere is called spaceflight.

atmospheric braking or **drag** or **friction** Resistance that slows a spacecraft as it enters and travels through Earth's atmosphere.

attitude The orientation of a spacecraft determined by the relationship between its horizontal, vertical, and lateral axes, the direction in which it is moving, and external reference points (usually stars but sometimes also the horizon).

attitude control thruster A small rocket engine fired to maneuver a space vehicle, such as an Apollo capsule. These thrusters are usually used in clusters.

Baikonur Cosmodrome A former Soviet space center in the Republic of Kazakhstan.

Bean, Alan L. (b. 1932) U.S. astronaut; Apollo 12 lunar module pilot and commander of Skylab 3; one of twelve astronauts to land on the Moon.

booster A rocket stage that powers a spacecraft during takeoff. A booster is sometimes jettisoned after the first several minutes of flight, once its fuel is exhausted. The word *booster* can also designate the entire launch vehicle of the spacecraft.

Frank Borman

Borman, Frank (b. 1928) U.S. astronaut; commander of Gemini VII and Apollo 8.

Brand, Vance (b. 1931) U.S. astronaut; flew on the Apollo-Soyuz mission, as well as three Space Shuttle missions.

burn The firing of a spacecraft engine, especially during flight.

capture trajectory The trajectory of a spacecraft headed for a planet or Moon that uses the gravity of the planet or Moon to help slow the craft and bring it into orbit. The spacecraft is, in effect, captured by the celestial body's gravity.

cargo bay The area of the Space Shuttle, behind the crew compartment and in front of the engines, used to carry the payload. Once the Shuttle is in orbit, double doors in the cargo bay open to expose the vehicle's radiators. A payload can then be deployed into space.

Carpenter, M. Scott (b. 1925) U.S. astronaut. He flew the Aurora 7 capsule in the Mercury 7 mission.

Carr, Gerald P. (b. 1932) U.S. astronaut; commander of Skylab 4.

Casper The name of the command module for Apollo 16.

celestial navigation Navigation by reference to the apparently unchanging stars to determine the position of a ship, aircraft, or space vehicle.

Centaur

Centaur U.S. rocket upper stage, the first to use a liquid-propellant booster. Centaurs have been combined successfully with first-stage Atlas and Titan rockets to launch space probes, military payloads, large communications satellites, and weather satellites.

centrifugal force The outward force experienced by a body constrained to move in a curved path, as a planet or a spacecraft in orbit. *See* **centripetal force.**

centripetal force The inward force that constrains a body, as a planet or a spacecraft, to move in a curved path. In orbital motion, the centripetal force is gravity. In circular orbital motion, centripetal force gives rise to an equal and opposite centrifugal force, each acting on the body to keep it in orbit.

Cernan, Eugene A. (b. 1934) U.S. astronaut; flew on the Gemini IX and Apollo 10 missions and commanded the Apollo 17 mission; one of twelve astronauts to land on the Moon.

Chaffee, Roger B. (1935–1967) U.S. astronaut; died in the fire that destroyed the Apollo 1 capsule on the launchpad.

Challenger The second Space Shuttle, first launched on April 4, 1983. *Challenger* flew nine successful missions. On its last mission, on January 28, 1986, the Shuttle's fuel tank exploded 73 seconds after takeoff, at an altitude of about 10 miles, in view of spectators and TV crews on the ground. All seven crew members died in the accident. *Challenger* was also the name of the Apollo 17 lunar module.

Chang Zheng *See* **Long March launch vehicle.**

Charlie Brown The name of the Apollo 10 command module.

Chinese rocket *See* **powder rocket.**

Clementine A small 500-pound (227 kg) unpiloted spacecraft, launched in January 1994, which entered lunar orbit the following February to make highly detailed images of the Moon's surface.

CM command module *(See box, p. 213).*

Michael Collins

Collins, Michael (b. 1930) U.S. astronaut; flew on the Gemini X mission and piloted the command module of the historic Apollo 11 lunar landing mission.

Columbia The first Space Shuttle. Its first orbital voyage took place on April 12, 1981. *Columbia* was also the name of the Apollo 11 command module.

coma A diffuse envelope of dust and gas that surrounds the nucleus of a comet.

combustion chamber The part of a rocket propulsion system consisting of a chamber in which the fuel and oxidizer are brought together and ignited, resulting in hot gases that expand and escape

through the nozzle to produce thrust.

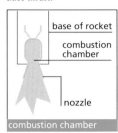
base of rocket
combustion chamber
nozzle
combustion chamber

comet A celestial body that consists of a head composed of a solid nucleus and a nebulous coma. In the part of its orbit close to the Sun, a comet develops a long vaporous tail pointing away from the Sun, frequently increasing its visibility from Earth.

comet

communications satellites Artificial satellites, often in geosynchronous orbit, that can relay television and radio signals, telephone calls, E-mail, data, and other modes of information almost instantly to and from different points on Earth. Some familiar communications satellites, past and present, are Echo, Telstar, Relay, Syncom, Intelsat, Westar, and Oscar.

Congreve rocket A gunpowder rocket developed in the early 1800s and used in the War of 1812, between the U.S. and Great Britain, as well as other battles. It was named after Sir William Congreve

(1772–1828), the British artillery expert who invented it. The rockets mentioned in "The Star-Spangled Banner" were Congreve rockets.

Conrad, Charles "Pete," Jr. (b. 1930) U.S. astronaut; flew on Gemini V and commanded Gemini XI, Apollo 12, and Skylab 2; one of twelve astronauts to land on the Moon.

Cooper, L. Gordon, Jr. (b. 1927) One of the seven original astronauts of Project Mercury, the first piloted spaceflight program of the United States. On the Mercury 9 mission, he orbited Earth 22 times in a capsule named *Faith 7*. Cooper also commanded Gemini V.

cosmic rays Charged particles of unknown origin traveling through space at close to the speed of light.

cosmonaut An astronaut trained in the Soviet Union or, after 1991, in Russia or the Republic of Kazakhstan.

countdown The audible counting backward in seconds from a starting number that marks the time remaining before a rocket or spacecraft launch. A typical countdown announcement would conclude, ". . . ten, nine, eight, seven, six, five, four, three, two, one: Liftoff."

crawler A huge, powerful, but slow tractor-tread vehicle used to transport a very large spacecraft, such as a

crawler

Space Shuttle or Saturn V launch vehicle, to the launchpad.

Crippen, Robert L. (b. 1937) U.S. astronaut; flew the Space Shuttle *Columbia* on its first mission and three missions aboard the Space Shuttle *Challenger* in 1983 and 1984.

cryogenic fuel A fuel stored in liquid state at extremely low temperature and often used for space missions because of its superior performance.

Cunningham, R. Walter (b. 1932) U.S. astronaut;

Apollo 7 crew member for the Earth-orbit checkout of the Apollo command and service modules.

deep space The regions of space well beyond Earth's atmosphere; interplanetary, interstellar, or intergalactic space. Also called *outer space*.

Deep Space Network (DSN) A network of three huge antennas, located at Goldstone, California, Madrid, Spain, and Canberra, Australia, that track the movements of spacecraft and satellites.

Delta A U.S. launch vehicle first developed in 1960,

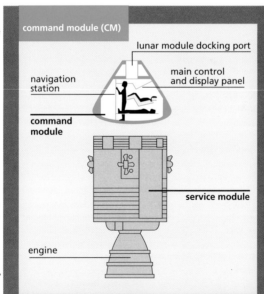

command module (CM)

lunar module docking port

navigation station

main control and display panel

command module

service module

engine

The section of the Apollo spacecraft containing the crew's living quarters and the instruments and equipment by which they controlled the flight and communicated with personnel back on Earth. The command module was about 11 feet (3.4 m) high and 13 feet (4 m) in diameter, with nearly four times more interior space than the Gemini module. It accommodated three astronauts. The main part of the module contained three couches for the astronauts, along with storage bays, food supplies, space suits, navigation systems, and sanitation facilities. Attached to the arm rests were various controls. Visibility was provided by five windows, two of them forward-facing to allow the astronauts to observe docking maneuvers.

Delta

derived from the Thor, with many design refinements over the years. Delta is now one of NASA's standard launch vehicles, serving as either a two-or three-stage rocket, as required.

Delta Clipper A single-stage-to-orbit launch vehicle under study by the U.S. Department of Defense and NASA.

descent engine A rocket on a spacecraft that by calculated firings enables the craft to make a controlled landing on a planetary or lunar surface.

descent stage A lunar module's lower section. It is equipped with legs and footpads for landing on the lunar surface, and it also serves as the launchpad for the ascent stage on its return to lunar orbit.

Discovery The third Space Shuttle, first launched on August 30, 1984. *Discovery* was the first Space Shuttle to fly after the 1986 *Challenger* accident.

dock (of two or more spacecraft) To join or become joined in space, especially so as to provide physical access from the interior of one spacecraft to that of another.

DSN *See* **Deep Space Network.**

Duke, Charles M., Jr. (b. 1935) U.S. astronaut; Apollo 16 lunar module pilot. He was one of twelve astronauts to land on the Moon.

E

Eagle The name of the lunar module for Apollo 11.

Early Bird The world's first commercial satellite and the first satellite of Intelsat. Also known as Intelsat 1, *Early Bird* was launched into geostationary orbit over the Atlantic Ocean by a Delta rocket in April 1965.

ERBS
(Earth resources satellite)

Earth resources satellites Artificial satellites that gather data for topographical and geological exploration, natural resources monitoring, and the like. Some Earth resources satellite programs are Landsat, SPOT, ERBS, and TOMS.

Echo I The first communications satellite. An inflated balloon with a metallized skin, launched in August 1960, *Echo I* bounced voice and TV signals from one ground station to another by simple reflection, marking the beginning of global communications service.

Eisele, Donn F. (b. 1930) U.S. astronaut. He was Apollo 7 command module pilot.

ELV *See* **expendable launch vehicle.**

Endeavour

Endeavour The fifth and newest Space Shuttle. First launched on May 7, 1992, *Endeavour* was used for the mission to repair the Hubble Space Telescope, in December 1993. *Endeavour* was also the name of the Apollo 15 command module.

Energiia Russia's most powerful launch vehicle, a huge, expendable launcher capable of lifting enormous payloads. First tested in May 1987, it can launch satellites and space stations into low Earth orbit. In 1988, Energiia carried the space shuttle *Buran* and released it within range of its orbit for a test flight.

Enterprise A test Space Shuttle that, in 1977, was lifted piggyback to an altitude of 22,000 feet (6,706 m) by a Boeing 747 and released to glide to a landing. The *Enterprise* is now in the collection of the Smithsonian Institution's National Air and Space Museum.

environmental control In a spacecraft environment, the maintenance of proper levels of oxygen, cabin pressure, temperature, humidity, ventilation, and water supplies and the elimination of carbon dioxide and other waste.

escape tower A framework mounted atop a spacecraft and supporting a solid rocket motor that, in an emergency, would fire and separate the capsule from the booster and parachute it into the ocean. Also called *safety tower*.

escape velocity The speed an object must attain to overcome the gravitational attraction of Earth or another celestial body. For Earth, escape velocity is about 25,000 miles per hour (40,300 km/h).

ET *See* **external tank.**

European Space Agency (ESA) A space organization with representatives from Austria, Belgium, Denmark, Finland, France, Germany, Ireland, Italy, the Netherlands, Norway, Spain, Sweden, Switzerland, and the United Kingdom. Beginning in 1975, European countries joined forces to develop their own launch program and now launch commercial communications and weather satellites. ESA and NASA have worked collaboratively on several projects, including Spacelab and the Hubble Space Telescope. ESA also develops and launches scientific spacecraft.

EVA *See* **extravehicular activity.**

Evans, Ronald E. (b. 1933-1990) U.S. astronaut; Apollo 17 command module pilot.

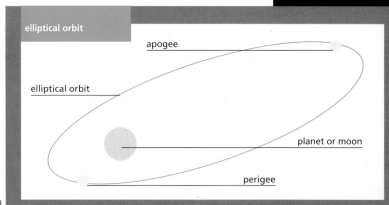

elliptical orbit

apogee

elliptical orbit

planet or moon

perigee

An orbit that describes the somewhat oval contour of an ellipse. The speed of a spacecraft in elliptical orbit will decrease as the craft approaches apogee—the point in

the orbit farthest from Earth—because the craft's orbital motion is going against the pull of Earth's gravity. As the spacecraft swings back and approaches its perigee—

the point in the spacecraft's orbit closest to Earth—it speeds up because its orbital motion and the planet's gravity are both pulling it in the same direction.

expendable launch vehicle (ELV) A rocket that can be used only once, as distinct from a reusable launch vehicle.

Explorer I The first successful American satellite. Explorer 1 was launched by a Jupiter C rocket on January 31, 1958, nearly four months after the Soviet Union launched Sputnik, the world's first artificial satellite.

external tank (ET) The Space Shuttle's huge expendable tank that contains liquid oxygen and liquid hydrogen propellants for the main engines.

extravehicular activity (EVA) Tasks and maneuvers, such as a space walk, performed by an astronaut in space, outside the spacecraft or on the Moon, sometimes with the assis-

tance of jet-propelled maneuvering devices.

Falcon The name of the Apollo 15 lunar module.

Freedom 7 The name of the Mercury capsule flown by Alan Shepard, the first American in space, on May 5, 1961.

free fall The condition of free motion in a gravitational field. A spacecraft coasts in space in free fall, and anything within the orbiting spacecraft that is not fastened will float as if there were no gravity. *See* **weightlessness.**

Friendship 7 The Mercury capsule in which John Glenn became the first American to orbit Earth.

Yuri Gagarin

Gagarin, Yuri A. (1934–1968) Soviet cosmonaut; on April 12, 1961, he became the first person to travel in space, circling Earth aboard Vostok 1.

Gamma-Ray Observatory (GRO) *See* **Great Observatories.**

gantry Scaffolding that houses a launch vehicle on the launchpad, allows the testing, servicing, and fueling of the launch vehicle, and gives the astronauts access to the spacecraft. The gantry is rolled aside before launch.

Garriott, Owen K. (b. 1930) U.S. astronaut; science pilot on Skylab 3 and mission specialist on the first Spacelab mission (Space Shuttle flight STS-9).

Gemini project A U.S. spaceflight program (1964–1966) designed to develop techniques needed to fly to the Moon. The focus of the twelve missions was to put two men in space, rendezvous and dock with another spacecraft, and have the astronauts walk in space. The Gemini successes laid the foundations for the Apollo program. Some of the significant missions were:

• **Gemini 3** (March 23, 1965) The first piloted test of a Gemini two-person spacecraft. Astronauts Grissom and Young flew the capsule *Molly Brown.*

• **Gemini IV** (June 3–7, 1965) A mission with astronauts McDivitt and White, notable for the first American space walk, carried out by White. Using a handheld thruster, he propelled himself for 22 minutes on his lifeline and practiced moving around while traveling nearly 18,000 miles per hour (29,000k/h) in orbit.

• **Gemini VI** (December 15–16, 1965) A mission marked by the first meeting of two piloted space vehicles, as astronauts Schirra and Stafford rendezvoused with Gemini VII.

• **Gemini VII** (December 4–18, 1965) Launched eleven days before Gemini VI, this mission, flown by astronauts Borman and Lovell, set a new record by staying in space for two weeks, a record unbroken until the Skylab missions of 1973–1974.

• **Gemini X** (July 18–21, 1966) During this mission astronauts Young and Collins docked with one Agena target and rendezvoused with another. Collins gathered scientific packages aboard the second Agena by exiting the spacecraft and propelling himself with a handheld thruster.

• **Gemini XII** (November 11–15, 1966) The final

mission of the series, during which astronauts Lovell and Aldrin rendezvoused with an Agena target vehicle and Aldrin set a Gemini duration record for a space walk of over five hours.

geophysical satellites Artificial satellites that perform services such as measuring, with great accuracy, distances between widely separated points on Earth and gathering data on phenomena in Earth's atmosphere and magnetic field. Two geophysical satellites are Lageos and OGO.

geostationary satellite or **geosynchronous satellite** A satellite in a geostationary orbit *(See box, p. 217).*

Gibson, Edward G. (b. 1936) U.S. astronaut; science pilot on Skylab 4.

GIRD An organization of engineers in Moscow and Leningrad in the 1930s who studied the principles of rocket propulsion.

Glenn, John H., Jr. (b. 1921) U.S. astronaut; one of the original seven astronauts of the Mercury program and the first American to orbit Earth. Glenn was launched in the space capsule *Friendship 7* by a Mercury-Atlas rocket on February 20, 1962, and circled Earth three times in five hours. Since 1974 he has been a U.S. senator from Ohio.

Goddard, Robert Hutchings (1882–1945) U.S. physicist; the first person to build liquid-propellant rockets that flew (1926)

Robert Goddard

and to build one that was able to break the sound barrier (1935). In 1919 the Smithsonian Institution published his small book on rocket propulsion, which mentions how rockets might be used to reach the Moon. He also foresaw the multistage rocket and rocket devices such as the nozzle and combustion chamber. NASA named the Goddard Space Flight Center near Washington, D.C., to honor him as the "Father of American Rocketry."

Gordon, Richard F., Jr. (b. 1929) U.S. astronaut; member of Gemini XI mission and Apollo 12 command module pilot.

gravity The attraction of the mass of a celestial body, such as Earth or the Sun, for other bodies. Gravity creates a sense of weight and causes objects to fall.

gravity-assisted trajectory A spacecraft's trajectory achieved by using the gravitational pull of a planet to redirect the spacecraft's course without expending fuel. Also called *slingshot orbit*.

Great Observatories Four large orbital observatories developed by NASA for astrophysical observa-

tion, each in a different part of the electromagnetic spectrum. Of these observatories, the Hubble Space Telescope and the Compton Gamma-Ray Observatory have already been launched. The Great Observatories are as follows:

• **Advanced X-ray Astrophysics Facility (AXAF),** designed to observe the X-ray part of the spectrum, will search for the presence of dark matter and will investigate hot young stars and galaxies, quasars, black holes, and other high-energy phenomena.

• **Compton Gamma-Ray Observatory (GRO),** designed to observe the most energetic part of the spectrum, studying pulsars, active galaxies, and quasars

and searching for evidence of antimatter.

• **Hubble Space Telescope** (HST), designed to observe galaxies in the far universe as well as objects in the solar system in visible and ultraviolet light.

• **Space Infrared Telescope Facility (SIRTF),** designed to observe the universe in the infrared part of the spectrum, investigating the birth of stars and the formation of planets and searching for dark matter.

Grissom, Virgil I. "Gus" (1926–1967) U.S. astronaut; piloted the second suborbital Mercury flight on July 21, 1961, in the capsule *Liberty Bell 7*, and in March 1965 commanded the Gemini 3 mis-

Virgil Grissom

geostationary orbit or geosynchronous orbit

satellite completes one orbit once every day

satellite travels 22,300 miles (or 35,888 kilometers) from Earth

satellite remains over one spot on Earth

An orbit in which a satellite moves at the same rate of spin as Earth, with an orbital period of 24 hours, so that the position of the satellite is fixed over a point on Earth and, when seen from Earth, seems to hover in nearly the same place in the sky. A satellite in geostationary orbit is at an altitude of approximately 22,300 miles (36,000 km) above the equator. Communications satellites are often placed in a geostationary orbit.

sion. He died in the fire that destroyed the Apollo 1 capsule on the launchpad.

Gumdrop The name of the Apollo 9 command module.

gyroscope A rotating disk or wheel mounted so that it can maintain a fixed orientation in space. Gyroscopes are used to stabilize aircraft, space vehicles, and rockets in flight.

HI and HII launch vehicles Japanese launch vehicles. The HI rocket is a three-stage launcher first developed in 1986 to launch unpiloted satellites into geostationary orbit. The HII rocket, a two-stage vehicle more powerful than the HI, was developed to launch larger unpiloted satellites, as well as lunar and planetary probes.

Fred Haise

Haise, Fred W., Jr. (b. 1933) U.S. astronaut; Apollo 13 lunar module pilot and one of four pilots for the Space Shuttle *Enterprise* test flights.

Halley's comet A famous comet named for the English astronomer Edmund Halley (1656–1742), the first person to predict the

return of a comet to the inner solar system in a regular orbit. Orbiting the Sun every 76 years, Halley's made its most recent appearance in 1986.

heat shield A barrier of protective material on the outside of a space vehicle that dissipates the intense heat of entry into an atmosphere, usually by vaporizing or melting.

heliocentric orbit An orbit that follows a path around the Sun.

Hubble Space Telescope *See* **Great Observatories.**

hybrid propellant A rocket propellant that is a combination of a solid fuel and a liquid oxidizer.

hybrid rocket A rocket motor that typically uses hybrid propellant.

hypergolic Of or relating to a propellant in which the fuel and oxidizer ignite spontaneously upon contact, without needing an igniter.

ICBM *See* **intercontinental ballistic missile.**

inertial navigation Navigation of a spacecraft or aircraft by means of gyroscopes or accelerometers, instruments which detect changes in direction and speed without reliance on an external frame of reference. Also called *inertial guidance*.

Inertial Upper Stage (IUS) A U.S. rocket booster used with a Titan launch vehicle or the Space Shuttle to launch heavy satellites into low Earth orbit or to send spacecraft beyond Earth orbit on interplanetary missions. The IUS consists of two stages and uses solid propellant.

intelligence satellites Artificial satellites containing electronic equipment for gathering data useful especially in intelligence or espionage, such as making images of secret military installations or listening in on radio and radar transmissions, and even private telephone conversations, on Earth. Also called *spy satellites*.

Intelsat (International Telecommunications Satellite Organization) An international organization formed by eleven countries in August 1964 to create and manage a worldwide system of communications satellites. The satellites are also called Intelsat.

Intelsat 1 *See* **Early Bird.**

intercontinental ballistic missile (ICBM) A long-range military rocket armed with a nuclear warhead and designed to follow a suborbital trajectory and reenter the atmosphere to descend on a target typically thousands of miles away. The Atlas rocket of the Mercury program and the Titan rocket of the Gemini program were ICBMs adapted for nonmilitary use.

intergalactic Located, occurring, or moving in space among the galaxies of the universe.

interplanetary Located, occurring, or moving in space among the planets of the solar system.

interstellar Located, occurring, or moving in space among stars, especially those within the Milky Way galaxy.

Intrepid The name of the Apollo 12 lunar module.

ionosphere The layer of Earth's atmosphere that begins at about 30 miles (48 km) above the surface and is dominated by charged, or ionized, atoms.

ion rocket A rocket designed to use high-speed, ionized (electrically charged) particles as its power source. It is still in the experimental stage.

Irwin, James B. (1930–1991) U.S. astronaut; Apollo 15 lunar module pilot and one of twelve astronauts to land on the Moon.

Jarvis, Gregory (1944–1986) A payload specialist from Hughes Aircraft Company who died in the Space Shuttle *Challenger* accident.

Jericho II An Israeli medium-range ballistic missile on which the design of the launch vehicle Shavit is based.

Jupiter C rocket An intermediate-range, three-stage ballistic missile developed by von Braun and his team for the U.S. Army. The Jupiter used liquid oxygen and kerosene as propellants and had a range of 1,500 miles (2,400 km). It was successfully launched into the atmosphere in 1956. The letter C in the rocket's name stands for "composite," in this case a combination of different rockets put together in three stages. A modified Jupiter C rocket, with a solid-propellant fourth stage designed to carry a satellite, launched the first U.S. satellite, *Explorer 1*, into orbit on January 31, 1958, nearly four months after the Soviet Union launched Sputnik, the world's first artificial satellite.

Kerwin, Joseph P. (b. 1932) U.S. astronaut and physician; science pilot on Skylab 2.

Kitty Hawk The name of the Apollo 14 command module.

Komarov, Vladimir M. (1927–1967) Soviet cosmonaut; commanded the Voskhod 1 in October 1964, the first three-person

Vladimir Komarov

orbital flight. Komarov died during a crash landing of the Soyuz 1 spacecraft in April 1967.

Sergei Korolëv

Korolëv, Sergei P. (1907–1966) Soviet space pioneer; directed the development of some of the first Soviet liquid-propellant rockets. He devoted his energies to rocketry from the 1920s on and became the director of GIRD. In 1946 he began work on long-range ballistic missiles, with the first Soviet two-stage ICBM being launched successfully in 1957. Korolëv adapted an ICBM as the rocket launcher of Sputnik, the world's first artificial satellite, and a year later helped launch Vostok 1, the first piloted spacecraft. He managed development of the Voskhod 1, the first three-person spacecraft to orbit Earth.

lander *See* **lunar module.**

Landsat A series of satellites in near-polar orbit that survey Earth in both the visible and the infrared spectrums and provide information for the assessment and management of Earth's natural resources, as well as valuable data about such matters as flood damage and oil spills. Landsat 1 was launched in 1972.

launch control center A facility at Cape Canaveral to control Apollo launches before transferring mission control to Houston. The center is now used for Space Shuttle launches.

launcher *See* **launch vehicle.**

launchpad A nonflammable platform used as a foundation from which launch vehicles and rockets are sent into space. Also called *launching pad.*

launch vehicle The rocket-powered system, made up of stages or sections, that boosts a space-

craft from Earth into space and gives it enough speed to achieve orbit. As each stage exhausts its fuel, it is jettisoned and falls back to Earth. Also called *launcher.*

launch window The calculated period of time in which a rocket must be launched if the mission is to meet its trajectory, rendezvous, or return requirements. The needs of the specific mission determine the exact duration of the launch window, which can be as brief as a few minutes.

LEM Lunar excursion module. (*See* **lunar module** box below).

Leonov, Aleksei A. (b. 1934) Soviet cosmonaut; the first man to walk

in space, on March 18, 1965, on the flight of the two-person Voskhod 2. Ten years later he was the commander of the Apollo-Soyuz Test Project.

Liberty Bell 7 The name of the capsule flown by Gus Grissom on the second Mercury flight. It sank shortly after splashdown.

liftoff The moment at the end of a countdown when a rocket ignites and launches skyward.

liquid propellants Chemical fuels, such as alcohol, kerosene, liquid hydrogen, and hydrazine, as well as oxidizers, such as nitrogen tetroxide and liquid oxygen. The fuel is the substance that burns, with the help of an oxidizer.

lunar module (LM)

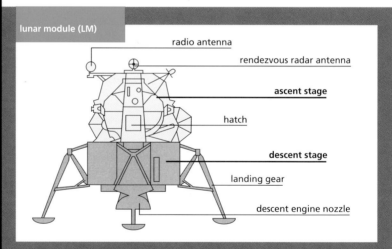

radio antenna
rendezvous radar antenna
ascent stage
hatch
descent stage
landing gear
descent engine nozzle

A spacecraft designed to carry astronauts from the Apollo command module to the lunar surface and back and to be their habitat on the Moon. Since it would not be required to fly anywhere but in the vacuum of space, designers of the lunar module could disregard aerodynamic rules and give the craft whatever shape was required. A lumpy, uneven surface would not have an impact on its flight. The lunar module is also sometimes known as a *lander* or a *lunar excursion module.*

LM *See* **lunar module** *(See box, p. 220).*

Long March launch vehicle Any of a family of two- and three-stage rockets used by China to launch unpiloted satellites into orbit. Since 1970 Long March rockets have been used to launch scientific, communications, weather, and military satellites. This family of rockets is also known by the Chang Zheng numbers CZ-1, CZ-2, CZ-3, and CZ-4. *Chang Zheng* means "long march" in Chinese.

Lousma, Jack R. (b. 1936) U.S. astronaut; Skylab 3 pilot and commander of the third Space Shuttle mission.

James Lovell

Lovell, James A., Jr. (b. 1928) U.S. astronaut; flew Gemini VII and Gemini XII; command module pilot on Apollo 8 and commander of Apollo 13.

low Earth orbit The orbit around Earth that a spacecraft makes when it is between one hundred and a few hundred miles above the surface.

Lucian (b. circa A.D. 120, d. after 180) Greek satirist whose tale *Vera Historia* ("The True History") is perhaps the first description of a journey to the Moon.

lunar Of, relating to, or involving the Moon.

M

Magellan An unpiloted spacecraft that was launched from the Space Shuttle *Atlantis* in May 1989 and reached Venus orbit 15 months later. Magellan carries radar instruments capable of penetrating the planet's thick cloud cover to image and map the surface. Its primary mission was completed in 1994.

magnetic field The region of space around a magnetic body, or one carrying electric current, within which charged particles are subjected to magnetic forces. Earth, the Sun, and many other celestial bodies have significant magnetic fields.

magnetosphere A region surrounding a celestial body, such as Earth or the Sun, where the magnetic field of the body is dominant and charged particles become trapped.

Mariner A NASA program of unpiloted planetary probes. Ten missions, Mariners 1 through 10, were launched from July 1962 to November 1973, exploring Mars, Venus, Mercury, and the interplanetary space environment.

Mattingly, Thomas K., II (b. 1936) U.S. astronaut. He was the Apollo 16 command module pilot and commander of two Space Shuttle missions.

McAuliffe, Christa (1948–1986) A high-school teacher from New Hampshire who died in the Space Shuttle *Challenger* accident. She was selected from eleven thousand applicants to be the first teacher in space.

James McDivitt

McDivitt, James A. (b. 1929) U.S. astronaut; commander of Gemini IV and Apollo 9.

McNair, Ronald E. (1950–1986) U.S. astronaut; died in the Space Shuttle *Challenger* accident.

Mercury project A U.S. space program (1961–1963) of piloted spaceflight. There were nine missions, six piloted. The names of all the capsules, each designed to carry one person, ended with the number 7, indicating the number of astronauts chosen for the project. The Mercury program gave the United States valuable experience in launching astronauts into space and returning them safely to Earth. The following two Mercury missions made especially significant contributions to space exploration:

• **Mercury 3** (May 5, 1991), on which Alan B. Shepard piloted the *Freedom 7* in a suborbital flight

to become the first American in space.

• Mercury 6 (February 20, 1962), on which John H, Glenn, piloting the *Friendship 7*, became the first American to orbit Earth.

meteorological satellites Artificial satellites that provide weather-pattern imagery and data on Earth temperatures, storm systems, atmospheric humidity, snow cover, and the like, as well as providing relays for distress signals. Some meteorological satellites are TIROS, NOAA, and GOES.

GOES (meteorological satellite)

microgravity A condition in which the noticeable effects of gravity are negligible; zero gravity or weightlessness.

Mir A Soviet space station launched on February 20, 1986, that is the successor to the Salyut space stations. Among Mir's significant design features are provision for more privacy and comfort for the occupants and a forward docking port allowing the docking of up to six modules.

mission control A NASA facility that monitors piloted spaceflights at the Johnson Space Center in Houston, Texas.

mission specialist A member of a Space Shuttle crew who is responsible for subsystems and payload activities.

Mitchell, Edgar D. (b. 1930) U.S. astronaut; Apollo 14 lunar module pi-

lot; one of twelve astronauts to land on the Moon.

Molly Brown The Gemini 3 capsule, named after the early-1960s Broadway musical *The Unsinkable Molly Brown*. It was the only Gemini capsule to be given a name.

MT Metric ton.

N

nanotechnology The technology of very small systems that may lead in the next century to the development of smaller, lighterweight satellites requiring smaller launch vehicles. Miniature satellites could possibly replace huge ones.

NASA (National Aeronautics and Space Administration) An agency of the U.S. government established by President Eisenhower in 1958 to implement space policy and direct the nation's efforts toward the scientific exploration and commercial uses of space.

navigation satellites Artificial satellites that provide data on the location of vessels and vehicles on Earth, including ones in distress, as well as information for oil exploration, marine fishing, and surveying. Some navigation satellite systems are Transit, Navstar/GPS, and COSPAS-SARSAT.

Newton, Sir Isaac (1642–1727) English mathematician and physicist who discovered the laws of gravity and motion.

Newton's third law of motion A statement in physics formulated by Isaac Newton: whenever there is an action on one body, there is an equal and opposite reaction on another body. For example, the force on the gases that leave a rocket engine in one direction equals exactly the thrust that propels the rocket in the opposite direction.

nozzle The rear part of a rocket engine through which the exhaust gases exit from the combustion chamber and accelerate to a high velocity, hurling the rocket skyward. The design of the nozzle affects a rocket's performance.

nuclear pulse propulsion A proposed use of nuclear energy to propel a rocket. The heat of the nuclear reaction would be converted into electrical power, which in turn would accelerate high-speed particles being ejected from the rocket. Also called *pulsed fusion propulsion*.

O

Hermann Oberth

Oberth, Hermann (1894–1989) German (Austrian-born) scientist of rocket theory and astronautics.

SEASAT (oceanographic satellite)

oceanographic satellites
Artificial satellites that gather data about Earth's seas and the marine environment. Oceanographic satellites include Seasat and TOPEX/Poseidon.

Odyssey The name of the Apollo 13 command module.

Ofeq 1 An Israeli satellite launched into low Earth orbit by the launch vehicle Shavit in September 1988.

OMS *See* **orbital maneuvering system.**

Onizuka, Ellison S. (1946–1986) U.S. astronaut; died in the Space Shuttle *Challenger* accident.

orbit The path described by one body as it circles another, such as that of Earth around the Sun, or an artificial satellite around Earth. *See* **elliptical orbit** *(See box, p. 215)*, **geostationary orbit** *(See box, p. 217)*, **heliocentric orbit, low Earth orbit, parking orbit, polar orbit, slingshot orbit, Sun-synchronous orbit, transfer orbit.**

orbital maneuvering system (OMS) A system consisting of two rocket en-

gines at the back of the Space Shuttle that thrust the Shuttle into and out of its orbit. The OMS is also used for major maneuvers while in orbit, such as rendezvous with another spacecraft or transfer to a higher or lower orbit.

orbital period The time required to complete one orbit.

Orbiter The piloted vehicle in the Space Shuttle system, the only part that goes into space. The solid rocket boosters and external tank used during launch are detached and jettisoned before the Orbiter enters its orbit.

O-ring seal A gasket sealing a joint, as between two sections in a Solid Rocket Booster. A rupture in the O-ring seals was found to have caused the Space Shuttle *Challenger* accident in January 1986.

Orion The name of the Apollo 16 mission lunar module.

oxidizer A propellant, such as liquid oxygen, that supports the combustion of a fuel in a rocket's combustion chamber. Also called *oxidant, oxidizing agent.*

parking orbit A temporary orbit for a satellite or a spacecraft.

payload Passengers, equipment, instruments, or

satellites that a spacecraft carries to complete its assigned mission, or the object placed in space by a launch vehicle.

payload bay *See* **cargo bay.**

payload specialist A member of a Space Shuttle crew who is not a NASA astronaut but who performs specialized duties with payloads or other activities unique to the specific mission; usually a scientist.

Peenemünde An early German rocket test site on the Baltic Sea where von Braun and other scientists developed the V1 and V2 military rockets.

Pegasus rocket

Pegasus rocket A U.S. three-stage, winged, solid-propellant rocket used to deliver small satellites into low Earth orbit. It is carried to high altitude by a jet aircraft and released, whereupon the rocket engine fires, at a safe distance from the aircraft, and the rocket continues into space. Named after the winged horse of Greek mythology, Pegasus was first launched in 1990. It is replacing the Scout as a small-payload launch vehicle.

Pegasus satellite Any of three satellites, unrelated to the Pegasus rocket, that were launched in 1965 by Saturn I rockets to detect meteoroids in near-Earth space.

perigee The point in the orbit of a satellite that is nearest Earth.

perturbation The effect that a body's gravity has on the orbit of another body. The attraction of the Moon, for example, causes Earth to weave slightly in its path around the Sun.

Pioneer A NASA program of unpiloted planetary probes. Thirteen missions were launched from October 1958 to August 1978. The first three missions were lunar probes that failed to reach the Moon but provided important data on the space environment. Pioneers 4 through 9 orbited the Sun and studied the inner solar system. Pioneer 10 flew close to Jupiter. Pioneer 11 reached Saturn and sent images and new information about the planet back to Earth. The last missions, Pioneer Venus 1 and 2, gathered data about Venus.

pitch Motion in a spacecraft about its lateral axis, so that the nose and tail move up and down.

Pogue, William R. (b. 1930) U.S. astronaut; Skylab 4 pilot.

polar orbit An orbit that follows a path over the poles of a celestial body, such as Earth or the Moon.

powder rocket The earliest known rocket, used in war by the Chinese and others as early as the eleventh century. In its simplest form it was a tube or rod filled with gunpowder. When lit, the gunpowder exploded and thrust the rocket forward to spread fire upon landing. Powder rockets were also used as fireworks. Also called *Chinese rocket*.

probe A spacecraft, such as Voyager or Galileo, sent forth to study celestial bodies and outer space. Probes make scientific observations and radio the data back to Earth.

propellant The combination of rocket fuel and oxidizer that a rocket engine burns to achieve thrust. The propellant typically constitutes about 90 to 95 percent of a rocket's weight and can be in either liquid or solid form or a hybrid.

propulsion system Any of various engine systems that deliver the propulsive force necessary to launch a vehicle or satellite into space. Spacecraft and satellites often have small, on board propulsion systems for in-space maneuvers.

Proton A powerful Russian launch vehicle used for putting massive space station modules into low Earth orbit, launching heavy satellites into geostationary orbit, and sending probes toward the inner planets. For years secrecy surrounded the Proton, first launched in 1965. This changed when Russia en-

tered the world space-launch market in 1988 to promote the commercial use of Proton—Russia claimed it could launch commercial satellites more cheaply than either NASA or ESA.

pulsed fusion propulsion *See* **nuclear pulse propulsion.**

quasar Any of a class of celestial objects lying at immense distances from the solar system and marked by their tremendous output of energy.

radiation Electromagnetic energy that includes gamma rays, x-rays, ultraviolet light, visible light, infrared light, microwaves, and radio waves. Radiation also refers to high-energy charged particles, such as protons, electrons, and ions, that are produced by the disintegration of radioactive nuclei or are naturally present in space in the form of solar wind particles and cosmic rays.

radiometer An instrument that measures the intensity of radiant energy and is used in meteorological satellites and planetary probes to obtain images that show relative surface temperatures.

Ranger A series of nine unpiloted NASA probes launched between August 1961 and March 1965 to study the Moon. Although some of the missions failed,

Ranger probe

the program was considered an overall success because of the high-quality images obtained of the lunar surface.

Reaction Control System (RCS) A Space Shuttle control system comprising three groups of small thruster engines located in the nose and on either side of the rear main engines. This system, consisting of a total of 44 thrusters, is used for small maneuvers, such as changing the spacecraft's orientation or final approach in a rendezvous with another spacecraft.

Redstone A U.S. rocket, designed in 1950 by Wernher von Braun and his group for the U.S. Army as a long-range, surface-to-surface missile. After the launch of Sputnik in 1957, the Redstone was redesigned as the first stage of the rocket used to launch the first U.S. satellite, Explorer 1, in 1958. Making only suborbital flights, a Redstone rocket launched Alan Shepard on his historic Mercury mission in 1961.

reentry The descent of a spacecraft into Earth's atmosphere from space.

rendezvous simulator

rendezvous The close approach of two spacecraft in orbit; sometimes culminates in docking of the two vehicles.

Resnick, Judith A. (1949–1986) U.S. astronaut; the

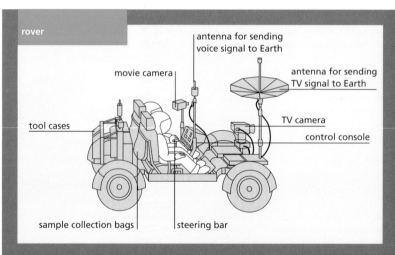

rover

- antenna for sending voice signal to Earth
- movie camera
- antenna for sending TV signal to Earth
- tool cases
- TV camera
- control console
- sample collection bags
- steering bar

A collapsible, battery-powered, four-wheeled vehicle used by astronauts to explore the lunar surface on Apollo missions 15, 16, and 17. The lunar rover looked a little like an elongated dune buggy and was capable of reaching 10 miles an hour (16 km/hr) on the lunar surface. Two 36-volt batteries powered the rover, and the vehicle weighed 460 pounds (209 kg). At the end of each mission on which a rover was used, the vehicle was left behind on the Moon when the astronauts returned to join the command module in orbit. More recently, the name rover has been applied to any mobile robotic vehicle whose intended use is the exploration of the surface of other planets or moons.

second U.S. woman astronaut in space and mission specialist on Space Shuttle *Discovery* in August 1984; died in the Space Shuttle *Challenger* accident.

retro-rocket pack

retro-rocket A rocket engine that slows a spacecraft or other vehicle by producing thrust in the direction opposite to the forward motion of the vehicle.

Ride, Sally K. (b. 1951) U.S. astronaut; the first American woman astronaut in space. She was mission specialist aboard the Space Shuttle *Challenger* in June 1983 and October 1984.

rocket A vehicle propelled by engines that deliver enough thrust to lift it beyond Earth's atmosphere and into space. A rocket works by controlled explosion or burning of propellants in its combustion chamber, moving in the opposite direction from the thrust of the escaping exhaust gases. Historically, the Chinese first used gunpowder rockets by the eleventh century as weapons to fight invaders. In the twentieth century, rockets called ballistic missiles were developed to carry sophisticated weapons. Today, rockets regularly carry satellites into space to gather and transmit information. They also launch piloted space-

craft into Earth orbit and robotic spacecraft on planetary missions.

roll Motion in a spacecraft in which the craft rotates around its longitudinal axis, or direction of flight.

Roosa, Stuart A. (b. 1933) U.S. astronaut; Apollo 14 command module pilot.

S

safety tower *See* **escape tower.**

Salyut A series of seven Soviet space stations. Salyut 1 was put into orbit on April 19, 1971. It demonstrated that space could be a habitable environment and led to the launches of the other stations in the Salyut series. Salyut means "salute" in Russian.

Sandal A Soviet ballistic missile that led to the design of the Cosmos rocket.

satellite A relatively small celestial body, such as the Moon, that orbits a larger body, such as the Earth; an artificial orbiting device, such as a communications or weather satellite.

Saturn rockets Three large and powerful launch vehicles, the Saturn I, Saturn IB, and Saturn V, were developed by NASA for the Apollo program and used from January 1964 until July 1975. The two-stage Saturn I rocket was used for the unpiloted Earth or-

bital flight tests of the Apollo module. Its final flights launched Pegasus meteoroid detection satellites. The Saturn IB rocket was a more powerful version of the Saturn I; it launched the first piloted Apollo mission, Apollo 7; delivered three crews to Skylab; and launched the Apollo-Soyuz mission, the last Saturn mission. The Saturn V was a massive, powerful three-stage launch vehicle with a reputation for reliability. It launched Apollo 9 and all nine of the missions to the Moon. A third stage of Saturn V became the Skylab space station, also launched by a Saturn V.

Savitskaya, Svetlana E. (b. 1948) Soviet cosmonaut; the second woman in space and the first woman to walk in space.

Schirra, Walter M. "Wally," Jr. (b. 1923) U.S. astronaut; one of the seven original astronauts of the Mercury program. In October 1962 he orbited Earth six times in a capsule named *Sigma 7,* and later flew on Gemini VI and commanded Apollo 7.

Schmitt, Harrison H. "Jack" (b. 1935) U.S. astronaut and geologist; Apollo 17 lunar module pilot; one of twelve astronauts to land on the Moon. He was later a U.S. senator from New Mexico.

Russell Schweickart

Schweickart, Russell L. "Rusty" (b. 1935) U.S. astronaut; Apollo 9 lunar module pilot.

Scobee, Francis R. "Dick" (1939–1986) U.S. astronaut; piloted a *Challenger* Space Shuttle mission in April 1984, died in the *Challenger* accident in January 1986.

Scott, David R. (b. 1932) U.S. astronaut; flew on Gemini VIII, command module pilot on Apollo 9, and commander of Apollo 15; one of twelve astronauts to land on the Moon.

Scout A lightweight, relatively inexpensive solid-propellant rocket developed in 1960. The smallest of the U.S. launch vehicles, it had three stages when first developed, and a fourth stage was added later. The Scout was used for more than one hundred successful launches until 1993, and has now been replaced by the Pegasus rocket.

second stage *See* **upper stage.**

service module The module of the Apollo spacecraft containing the main engine, fuel cells, water, and other systems and supplies.

Shavit launch vehicle Israel's two-stage solid-propellant rocket first used in September 1988 to launch its first satellite Ofeq I into low Earth orbit. Shavit means "comet" in Hebrew.

Alan Shepard

Shepard, Alan B., Jr. (b. 1923) U.S. astronaut; one of the original seven astronauts of the Mercury program and the first American in space. The historic Mercury mission occurred on May 6, 1961, when a Redstone rocket launched Shepard on a 15-minute suborbital flight in a one-person, bell-shaped space capsule he named *Freedom 7*. Shepard's flight confirmed that the Mercury capsule design was sound. He later commanded the Apollo 14 mission and was one of twelve astronauts to land on the Moon.

Skylab A space station assembled from Saturn and Apollo parts and launched into orbit on May 14, 1973. It served as a laboratory for scientific experiments in space until February 1974. In 1979, it reentered the Earth's atmosphere; most of it burned up but some parts fell harmlessly into the In-

dian Ocean and onto Australia. Skylab was the world's first big orbital workshop-laboratory. The missions associated with Skylab are as follows:

Skylab 1 A mission begun on May 14, 1973, in which an unpiloted space station made from one stage of a Saturn V rocket was launched. This orbiting laboratory was occupied by three separate crews, flown up to live and conduct experiments there.

Skylab 2 This mission (May 25–June 22, 1973) was Skylab's first piloted mission, with astronauts Conrad, Kerwin, and Weitz as crew. In a three-and-one-half-hour space walk, they first repaired serious damage to the orbital workshop incurred during its launch and then conducted scientific and medical experiments. This successful mission set a new space endurance record of 28 days.

Skylab 2

Skylab 3 This was the second piloted Skylab mission (July 28–September 25, 1973). Remaining in space for 59 days, astronauts Bean, Garriott, and Lousma continued scientific and medical experiments and observations of the Earth and Sun from orbit.

Skylab 4 This mission (November 16, 1973–February 8, 1974) was the third and final visit of a crew to the Skylab space station. Astronauts Carr, Gibson, and Pogue obtained medical information about themselves to further the study of extended piloted spaceflight. Members of the crew conducted the longest U.S. space mission to date.

Slayton, Donald K. "Deke" (1924–1993) U.S. astronaut; one of the original seven astronauts chosen for the Mercury program, he flew the Apollo-Soyuz Test Project.

slingshot orbit *See* **gravity-assisted trajectory.**

SLV3 launch vehicle India's first successful launch vehicle, a four-stage, solid-propellant rocket, launched its first satellite, Rohini, into low Earth orbit in July 1980. SLV stands for Satellite Launch Vehicle.

Smith, Michael (1945–1986) U.S. astronaut; died on his first mission, in the Space Shuttle *Challenger* accident.

Snoopy The name of the Apollo 10 lunar module.

solar panel An array of light-sensitive cells on a spacecraft, such as a satellite or probe, that is used to generate electrical power for the craft in space.

solar wind The flow of high-energy charged particles, such as electrons, protons, and ions, radiating outward from the Sun throughout the solar system. It reaches well beyond the planet Pluto and moves as fast as 450 miles per second (720 km/s).

solid propellant A chemical rocket propellant used in solid rather than liquid or gaseous form. When a block of solid propellant ignites in the core of a rocket's combustion chamber, the force of its exhaust propels the rocket. The term *solid propellant* refers to both the fuel and the oxidizer.

solid rocket booster (SRB) Either of two solid-propellant rockets that augment the main engines of the Space Shuttle, providing extra thrust during liftoff. After the first two minutes of flight, they are jettisoned, parachuted to Earth, and recovered from the sea. They are then refitted for later flights.

sounding rocket A rocket that travels in a suborbital trajectory to probe the upper atmosphere at different levels for information about atmospheric or near-space conditions. Sounding rockets have also been used to gather astronomical data, as during eclipses, although their observational period is brief.

Soyuz

Soyuz A Soviet launch vehicle made up of a rocket module, an orbital module, a docking assembly, and a descent module. Although the Soyuz was originally intended for piloted lunar expeditions, it has been used for rendezvous in Earth or-

bit. More than seventy missions have been flown since the program began in 1967. Since the launch of the first Salyut space station in 1971, most Soyuz spacecraft have been used to travel to Salyut stations and back. The Soyuz is still in use.

space The boundless region beyond Earth's atmosphere, between the planets, or beyond the solar system. Although it seems empty compared to Earth's atmosphere, space actually contains trace amounts of atoms and molecules, dust, and particles.

Space Age The modern historical period begun with the launch of Sputnik in 1957 and continuing to the present. It has been marked by the exploration of the solar system by piloted and unpiloted spacecraft, artificial satellites, and orbiting space telescopes, the latter also contributing to exploration of the universe beyond the solar system.

space capsule A pressurized module for carrying astronauts on spaceflights; for example, a piloted vehicle like those used in the Mercury program.

spacecraft A vehicle designed to travel into space beyond Earth's atmosphere, with or without a crew. Also called *spaceship* or *space vehicle.*

spaceflight Flight by means of spacecraft, in space beyond Earth's atmosphere.

Space Infrared Telescope Facility (SIRTF) *See* **Great Observatories.**

Spacelab A modular laboratory carried in the payload bay of the Space Shuttle's Orbiter. It can be configured for research in many scientific fields. Its first mission was in 1983. The European Space Agency developed Spacelab as part of the U.S. Space Transportation System.

space medicine The study of the physiological effects of microgravity and other spaceflight conditions on the human body.

spaceport An installation for testing, maintaining, and launching spacecraft.

Space Race A long-running competition between the United States and the Soviet Union that began during World War II, with the two nations striving to develop rockets. When the Soviets test-fired experimental missiles in the mid-1950s, the Americans intensified their program. Both countries then began programs to launch a satellite into space. The Soviets got there first with Sputnik, on October 4, 1957, and in 1961 they launched the first piloted spacecraft, Vostok 1. Confidence shaken, the United States made an all-out effort to catch up and take a decisive lead. That exertion paid off handsomely when the space mission Apollo 11 placed American astronauts on the Moon. Well before the collapse of the Soviet system in the late

1980s, the Space Race had ended, and competitiveness had turned into cooperation, not only between the United States and Russia, but among other nations, as well.

Space Radar Laboratory A laboratory put into orbit from the Space Shuttle *Endeavour* in April 1994 to gather data about the environment.

Space Shuttle main engines The three reusable main rocket engines at the back of the Orbiter that are fired during launch and operate until nine minutes after liftoff, when the orbital maneuvering system takes over. The main engines draw their propel-

lants from the huge external tank mounted under the Orbiter.

Space Shuttle program NASA's piloted spaceflight program after the Apollo program. Work began on the Space Shuttle in 1972, and the first flight took place in April 1981. The versatile Space Shuttle can be used as a laboratory to perform scientific experiments, as well as to deploy or repair commercial, defense, and scientific satellites and other instruments. The five U.S. Space Shuttles—*Columbia, Challenger, Discovery, Atlantis,* and *Endeavour*—have flown more than sixty flights since the first one in 1981.

Space Shuttle

external tank
solid rocket booster
orbiter
main engines

A vehicle that consists of an orbiter, two solid rocket boosters, and an external tank, a Shuttle is launched into orbit like a rocket and glides to a landing in the manner of an airplane. The Orbiter is what many people picture when they think of the Space Shuttle, not realizing that the name includes all three parts. The world's first reusable space transportation system, the Shuttle ferries people and payloads to low Earth orbit. It also operates in space as a laboratory and workshop.

space sickness Temporary dizziness and nausea sometimes affecting people early in a spaceflight.

space station A large satellite designed to remain in orbit around Earth for extended periods of time and to serve as living and working quarters for a crew who use it as a base, especially for scientific experiments and observations.

space suit A protective garment with self-contained life-support and communication systems allowing the wearer to survive and function in space.

space vehicle *See* **spacecraft.**

space walk Activity by an astronaut outside the spacecraft or on the surface of the Moon; extravehicular activity. A protective space suit must be worn during a space walk. *See* **extravehicular activity.**

Spider The name of the Apollo 9 lunar module.

splashdown The landing in the ocean of a spacecraft returned from orbit.

Sputnik 2

Sputnik Any of three Soviet satellites. The first,

launched October 4, 1957, was the first artificial satellite in space.

spy satellites *See* **intelligence satellites.**

SRB *See* **solid rocket booster.**

Thomas Stafford

Stafford, Thomas P. (b. 1930) U.S. astronaut; flew Gemini VI and Gemini IX. He was commander of Apollo 10 and the Apollo-Soyuz Test Project.

staging The use of two or more stages (rocket modules). Rocket stages are used to boost spacecraft to higher altitudes or to carry larger payloads. Most space rockets have three stages that function sequentially and are discarded once they have served their purpose. The primary job of the first stage, the bottom part of the rocket, is to achieve liftoff from the launchpad and carry the vehicle to a certain velocity and altitude. The second (upper) stage takes over once the fuel of the first stage is exhausted and boosts the payload to higher altitude or into orbit. A third or fourth stage would take over in similar fashion.

suborbital trajectory A trajectory followed by a rocket that does not go fast

enough to attain orbit and instead falls back to Earth in a long curve.

Sullivan, Kathryn D. (b. 1951) U.S. astronaut; became, in October 1984, the first U.S. woman astronaut to walk in space, while serving as mission specialist on the Space Shuttle *Discovery.*

Sun-synchronous orbit A near-polar orbit that passes over the same spot on Earth at the same local time every day.

Surveyor A NASA series of unpiloted lunar spacecraft. The program's goal was to land an unpiloted spacecraft on the Moon and gather data for the upcoming Apollo program. Seven missions were flown from May 1966 to January 1968. The successful Surveyor program demonstrated that the Moon's surface could support the Apollo lunar module without its sinking into the lunar dust.

Swigert, John L., Jr. (1931–1982) U.S. astronaut; Apollo 13 command module pilot.

Syncom 1 The first geostationary satellite, launched in February 1963.

T

telecommunications satellites *See* **communications satellites.**

Tereshkova, Valentina V. (b. 1937) Soviet cosmo-

Valentina Tereshkova

naut. She piloted Vostok 6, in June 1963, to become the first woman to fly in space.

Thor Intermediate-range ballistic missile developed by the U.S. Air Force by 1957. It used liquid oxygen and kerosene as propellants and had a range of 1,500 miles (2,400 km). Thor was combined successfully with Agena and Delta rockets. One of NASA's "workhorse" launch vehicles, Delta was based on the earlier Thor missile.

thrust The rearward force that is produced by exhaust gases escaping through a rocket engine. This force propels the vehicle upward or forward.

TIROS (Television Infrared Observation Satellite) A series of U.S. weather satellites, the first of which was launched in April 1960. Tiros 1 was the first satellite to send back pictures of cloud formations. Between 1960 and 1965 ten Sun-synchronous TIROS satellites were launched successfully. These were followed by two more generations of TIROS satellites, launched from 1966 to 1976, and yet another series from 1978 to 1981. The Department of Defense and NASA jointly

developed the TIROS, first for military purposes and later for civilian applications.

Titan A U.S. two-stage ICBM, developed by the Air Force and first tested in 1959. A series of Titan ICBMs became part of the U.S. strategic nuclear arsenal but have since been retired. Although they were not originally intended to launch satellites, they launched the Gemini missions in the 1960s and became a popular launch

Titan-Centaur

vehicle for civilian satellites. Titans have also been combined with Agena and Centaur upper stages. The Titan-Centaur (known as the Titan IIIE) has launched solar probes, two Viking spacecraft to Mars in 1976, and two Voyager spacecraft to the outer planets in 1977. The latest Titan launch vehicle is the largest, most powerful U.S. expendable launch vehicle in use. Its primary function is putting military satellites into orbit

Titov, Gherman S. (b. 1935) Soviet cosmonaut. The second man in space, he piloted Vostok 2 on August 6 and 7, 1961, orbiting Earth for 25 hours.

Gherman Titov

track To plot or observe the path of (a spacecraft, artificial satellite, or missile) with instruments.

trajectory The path or curve described by a body, such as a planet, rocket, or bullet, as it moves through space.

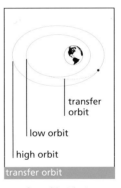
transfer orbit

transfer orbit The intermediate orbit that a satellite makes after it leaves low Earth orbit and before it reaches geostationary, lunar, or planetary orbit. Also called *Hohmann's transfer orbit* or *geostationary transfer orbit.*

transfer orbit stage (TOS) A launch vehicle upper stage designed to boost heavy payloads into geostationary transfer orbit.

Tsiolkovsky, Konstantin Eduardovich (1857–1935) Russian scientist considered

Konstantin Tsiolkovsky

to be the father of Soviet astronautics. He was the first to sketch and describe liquid-fuel rocket engines and a multistage rocket propulsion system. He realized that a rocket carries its own oxygen and therefore could operate in space, and that the principle of reaction propulsion (rocket motion) can operate in the vacuum of space. In his writings he also argued that liquid fuels could have more driving power than solid fuels like gunpowder, and could be more easily controlled. He wrote about spaceships, space suits, and colonizing space. A self-taught pioneer of spaceflight, he was the first to realize that it was actually possible for artificial satellites and space stations to orbit Earth.

U
V

upper stage The self-propelled stage of a launch vehicle that ignites after the first stage has fallen away and that carries the payload to be boosted into orbit.

V2 The first large, liquid-propellant rocket. Originally called the A4, the V2 was developed by Wernher von Braun for use by the German army during

World War II. In August 1942 an A4 test rocket broke the sound barrier. Germany fired thousands of V2s at London and other cities during the war, causing thousands of civilian casualties. The United States and the USSR captured a number of V2 rockets in Germany; after the war the U.S. army used its share to study the operation of large liquid-propellant rockets and to study the upper atmosphere. One V2 at that time reached an altitude of 114 miles (182 km).

Valier, Max (1895–1930) Austrian spaceflight popularizer and inventor who built cars propelled by rockets. His experiments stimulated general public interest in the possible uses of rockets. He was killed in an explosion during a test of a rocket-propelled car.

Van Allen belts Two doughnut-shaped belts of intense radiation (charged particles from the Sun) trapped in Earth's magnetic field. Named after James A. Van Allen (b. 1914), U.S. physicist, whose instrument on Explorer I first detected this radiation.

Vehicle Assembly Building (VAB) A 52-story building at the Kennedy Space Center that was constructed to house the Saturn V launch vehicle and the Apollo spacecraft on top of it. It is now used for the Space Shuttle.

Verne, Jules (1828–1905) French writer regarded as a founder of modern science fiction. Some of his writings

were amazingly prophetic of the science of the future. His novels include *Journey to the Center of the Earth* (1864), *From the Earth to the Moon* (1865), *Trip Around the Moon* (1870), and *Twenty Thousand Leagues Under the Sea* (1870).

vernier engine A small rocket engine used to adjust the trajectory or attitude of a large rocket. Named after the French mathematician Pierre Vernier.

Wernher von Braun

von Braun, Wernher (1912–1977) U.S. (German-born) rocket scientist. He directed the technical development of the V2 (A4) ballistic missile for the German army during World War II. After the war he came to the United States, eventually leading the team that in January 1958 launched America's first satellite, Explorer 1. Von Braun was founding director of NASA's Marshall Space Flight Center, in Alabama, which developed the Saturn rockets used during the Apollo program. He retired from NASA in 1972.

von Opel, Fritz (1899–1971) German industrialist and automaker who experimented with Max Valier on

Fritz von Opel

rocket propulsion for automobiles and aircraft. On September 30, 1929, he became the second person to fly a rocket-powered plane, the Opel-Sander Rak 1, covering nearly two miles in 75 seconds.

Voskhod 1 The Soviet space capsule piloted on a one-day mission on October 12, 1964, by Komarov, Feoktistov and Yegorov.

Voskhod 2 The Soviet space capsule flown in March 1965, during which cosmonaut Leonov achieved the first space walk in history.

Vostok The first operational Soviet ICBM, it was used in 1959 to launch the space probe Luna, the world's first satellite to escape Earth's gravity. Vostok was also the name of a series of six Soviet space capsules, the first of which, Vostok 1, was flown by cosmonaut Gagarin in history's first piloted spacecraft mission, in 1961. Vostok 2 was flown later that year.

W

weather satellites *See* **meteorological satellites.**

weightlessness The absence of noticeable effects of gravity. Astronauts feel weightless when floating in space or when they are inside a spacecraft that is undergoing the same acceleration of gravity as they are, as in orbit around Earth. Gravity pulls on both the astronauts and the spacecraft, keeping them in the same orbit, but the astronauts cannot feel the gravitational pull because there is no fixed surface, like the ground, to resist the acceleration. *See* **zero-gravity.**

weightlessness

Weitz, Paul J. (b. 1932) U.S. astronaut; Skylab 2 pilot and commander of the first flight of Space Shuttle *Challenger*.

Wells, H. G. (Herbert George) (1866–1946) British writer and historian whose science fiction novels, like those of Jules Verne, anticipated scientific discoveries of later times. His novels include *The Time Machine* (1895), *The War of the Worlds* (1898), and *The First Men in the Moon* (1901).

White, Edward H., II (1930–1967) U.S. astronaut; the first American to walk in space, during the Gemini IV mission of June 1965. He died in the fire that destroyed the Apollo 1 capsule on the launchpad.

Edward White

window *See* **launch window.**

Worden, Alfred M. (b. 1932) U.S. astronaut; Apollo 15 command module pilot.

X Y Z

Yankee Clipper The name of the Apollo 12 command module.

yaw Motion of a spacecraft about its vertical axis, so that the nose and tail move from side to side.

Young, John W. (b. 1930) U.S. astronaut. He flew on Gemini 3 and Gemini X and also was command module pilot on the Apollo 10 mission and commander of Apollo 16. He commanded Space Shuttle *Columbia's* first mission in April 1981 and the first Spacelab mission in November 1983. Young was one of twelve astronauts to land on the Moon.

zero-gravity A state of apparent weightlessness that occurs when an astronaut or a spacecraft is floating in space. Also called *zero g*. *See* **weightlessness.** *See also* **free fall** *and* **microgravity.**

Piloted Spaceflight

Program	Launch Date	Crew	Notable Events
Vostok 1	April 12, 1961	Gagarin	First human in space, orbital flight
Mercury 3	May 5, 1961	Shepard	First American in space; suborbital flight
Mercury 4	July 21, 1961	Grissom	Capsule sank but astronaut rescued
Vostok 2	August 6, 1961	Titov	First human in space for over 24 hours
Mercury 6	February 20, 1962	Glenn	First American in Earth orbit
Mercury 7	May 24, 1962	Carpenter	First meal eaten in space; missed landing site by 250 miles (402 km)
Vostok 3	August 11, 1962	Nikolayev	First joint mission (with Vostok 4)
Vostok 4	August 12, 1962	Popovich	Came within 4 miles (6.5 km) of Vostok 3
Mercury 8	October 3, 1962	Schirra	First splashdown in Pacific Ocean
Mercury 9	May 15, 1963	Cooper	First U.S. flight to exceed 24 hours
Vostok 5	June 14, 1963	Bykovsky	Second joint mission (with Vostok 6)
Vostok 6	June 16, 1963	Tereshkova	First woman in space
Voskhod 1	October 12, 1964	Komarov Feoktistov Yegorov	First 3-person mission
Voskhod 2	March 18, 1965	Belyayev Leonov	First extravehicular activity (EVA) completed by Leonov (20 minutes)
Gemini 3	March 23, 1965	Grissom Young	First U.S. 2-person crew; first manual maneuvering
Gemini IV	June 3, 1965	McDivitt White	First U.S. EVA by White (22-minute)
Gemini V	August 21, 1965	Cooper Conrad	8-day flight
Gemini VII	December 4, 1965	Borman Lovell	14-day flight; rendezvous with Gemini VI
Gemini VI	December 15, 1965	Schirra Stafford	Came within 12 inches (30 cm) of Gemini VII

Piloted Spaceflight

Program	Launch Date	Crew	Notable Events
Gemini VIII	March 16, 1966	Armstrong Scott	First docking (with Agena target vehicle)
Gemini IX-A	June 3, 1966	Stafford Cernan	2-hour space walk; most accurate splashdown
Gemini X	July 18, 1966	Young Collins	First dual rendezvous (with Agena 10, then Agena 8)
Gemini XI	September 12, 1966	Conrad Gordon	Highest altitude of 850 miles (1,368 km); first tethered flight
Gemini XII	November 11, 1966	Lovell Aldrin	Aldrin completed 5-hour EVA
Apollo 1	January 27, 1967	Grissom White Chaffee	Crew died in spacecraft fire during countdown test
Soyuz 1	April 23, 1967	Komarov	Cosmonaut died during reentry
Apollo 7	October 11, 1968	Schirra Eisele Cunningham	First live TV broadcast from piloted spacecraft
Soyuz 3	October 26, 1968	Beregovoy	Maneuvered near unpiloted Soyuz 2
Apollo 8	December 21, 1968	Borman Lovell Anders	First humans to orbit Moon; set speed record of 24,200 miles per hour (38,938 km/hr)
Soyuz 4	January 14, 1969	Shatalov	Docked with Soyuz 5; received 2 cosmonauts
Soyuz 5	January 15, 1969	Volynov Yeliseyev Khrunov	Transferred 2 crew members to Soyuz 4
Apollo 9	March 3, 1969	McDivitt Scott Schweickart	First flight test of lunar module (LM) in Earth orbit
Apollo 10	May 18, 1969	Stafford Young Cernan	Tested LM in lunar orbit 8.9 miles (14.3 km) from the Moon's surface
Apollo 11	July 16, 1969	Armstrong Collins Aldrin	First human landing on Moon

Piloted Spaceflight

Program	Launch Date	Crew	Notable Events
Soyuz 6	October 11, 1969	Shonin Kubasov	Joint mission with Soyuz 7 and 8 without docking
Soyuz 7	October 12, 1969	Filipchenko Gorbatko Volkov	
Soyuz 8	October 13, 1969	Shatalov Yeliseyev	
Apollo 12	November 14, 1969	Conrad Gordon Bean	Retrieval of pieces of Surveyor 3 spacecraft
Apollo 13	April 11, 1970	Lovell Swigert Haise	Deep space abort due to explosion in service module (SM); LM used as lifeboat; crew returned safely
Soyuz 9	June 1, 1970	Nikolayev Sevastyanov	Endurance record of nearly 18 days in space
Apollo 14	January 31, 1971	Shepard Roosa Mitchell	Third lunar landing; Shepard hit two golf balls on the Moon
Soyuz 10	April 22, 1971	Shatalov Yeliseyev Rukavishnikov	Docked with Salyut space station but did not board
Soyuz 11	June 6, 1971	Dobrovolsky Volkov Patsayev	Crew spent 22 days aboard Salyut; crew died during reentry
Apollo 15	July 26, 1971	Scott Worden Irwin	First use of lunar rover
Apollo 16	April 16, 1972	Young Mattingly Duke	Highest elevation for a lunar landing site at 25,688 feet (7,809 m)
Apollo 17	December 7, 1972	Cernan Evans Schmitt	Final Apollo lunar mission; nearly 75 hours spent on lunar surface
Skylab 2	May 25, 1973	Conrad Kerwin Weitz	Docked with Skylab space station for 28 days
Skylab 3	July 28, 1973	Bean Garriott Lousma	Docked for over 59 days; conducted 19 student experiments

Piloted Spaceflight

Program	Launch Date	Crew	Notable Events
Soyuz 12	September 27, 1973	Lazarev Makarov	Testing of improved Soyuz
Skylab 4	November 16, 1973	Carr Gibson Pogue	Last Skylab mission; astronauts traveled 34.5 million miles (55.5 million km); set U.S. space record of 84 days
Soyuz 13	December 18, 1973	Klimuk Lebedev	Conducted various astrophysical and biological experiments
Soyuz 14	July 3, 1974	Popovich Artyukhin	Docked with Salyut 3; crew boarded space station
Soyuz 15	August 26, 1974	Sarafanov Demin	Rendezvoused with Salyut 3; did not dock
Soyuz 16	December 2, 1974	Filipchenko Rukavishnikov	Testing for Apollo-Soyuz Test Project
Soyuz 17	January 10, 1975	Gubarev Grechko	Docked with Salyut 4; boarded station
Soyuz 18	May 24, 1975	Klimuk Sevastyanov	Docked with Salyut 4; boarded station
Soyuz 19 **Apollo 18**	**July 15, 1975** **July 15, 1975**	**Leonov** **Kubasov** **Stafford** **Slayton** **Brand**	**Apollo-Soyuz Test Project: first U.S.-USSR joint mission; docked Apollo 18 and Soyuz 19; crews shared meals**
Soyuz 21	July 6, 1976	Voynov Zholobov	Docked with Salyut 5; boarded space station
Soyuz 22	September 15, 1976	Bykovsky Aksenov	Conducted earth resources study
Soyuz 23	October 14, 1976	Zudov Rozhdestvensky	Unsuccessful docking with Salyut 5
Soyuz 24	February 7, 1977	Gorbatko Glazkov	Docked with and boarded Salyut 5
Soyuz 25	October 9, 1977	Kovalenok Ryumin	Unsuccessful docking with Salyut 6
Soyuz 26	December 10, 1977	Romanenko Grechko	Crew spent 96 days aboard Salyut 6; returned in Soyuz 27

Piloted Spaceflight

Program	Launch Date	Crew	Notable Events
Soyuz 27	January 10, 1978	Dzhanibekov Makarov	Spent nearly 6 days on Salyut 6; crew returned in Soyuz 26
Soyuz 28	March 2, 1978	Gubarev Remek	Docked with Salyut 6; first space traveler not from the U.S. or USSR
Soyuz 29	June 15, 1978	Kovalenok Ivanchenkov	Crew spent 139 days aboard Salyut 6; returned in Soyuz 31
Soyuz 30	June 27, 1978	Klimuk Hermaszewski	Docked with Salyut 6; stayed seven days
Soyuz 31	August 26, 1978	Bykovsky Jaehn	Crew spent almost 8 days on Salyut 6; returned in Soyuz 29
Soyuz 32	February 25, 1979	Lyakhov Ryumin	Crew spent 175 days on Salyut 6; returned in Soyuz 34
Soyuz 33	April 10, 1979	Rukavishnikov Ivanov	Unsuccessful docking with Salyut 6
Soyuz 35	April 9, 1980	Popov Ryumin	Crew spent nearly 185 days aboard Salyut 6; returned in Soyuz 37
Soyuz 36	May 26, 1980	Kubasov Farkas	Docked with and spent nearly 8 days aboard Salyut 6; returned in Soyuz 35
Soyuz T-2	June 5, 1980	Malyshev Aksënov	First piloted flight of newly developed Soyuz T
Soyuz 37	July 23, 1980	Gorbatko Pham	Crew included first Vietnamese cosmonaut
Soyuz 38	September 18, 1980	Romanenko Tamayo-Mendez	First black cosmonaut, Tamayo-Mendez, from Cuba
Soyuz T-3	November 27, 1980	Kizim Makarov Strekalov	First Soviet crew of three since 1971
Soyuz T-4	March 12, 1981	Kovalenok Savinykh	Docked with and boarded Salyut 6 for seven-day stay
Soyuz 39	March 22, 1981	Dzhanibekov Gurragcha	Crew included first Mongolian crew member in space
Columbia (STS-1)	**April 12, 1981**	**Young Crippen**	**First piloted orbital test flight of U.S. Space Shuttle**

Piloted Spaceflight

Program	Launch Date	Crew	Notable Events
Soyuz 40	May 14, 1981	Popov Prunariu	Seven-day visit to Salyut 6; last flight of Soyuz
Columbia (STS-2)	November 12, 1981	Engle Truly	Payload included OSTA 1; crew tested remote manipulator arm
Columbia (STS-3)	March 22, 1982	Lousma Fullerton	First use of alternate shuttle landing site
Soyuz T-5	May 13, 1982	Berezovoy Lebedev	Crew remained on Salyut 7 for 211 days; returned in Soyuz T-7
Soyuz T-6	June 24, 1982	Dzhanibekov Ivanchenkov Chrétien	Mission to Salyut 7
Columbia (STS-4)	June 27, 1982	Mattingly Hartsfield	Carried U.S. Department of Defense payload
Soyuz T-7	August 19, 1982	Popov Serebrov Savitskaya	Savitskaya became second Soviet woman in space
Columbia (STS-5)	November 11, 1982	Brand Overmyer Allen Lenoir	Launched commercial satellites (SBS 3 and Anik C-3)
Challenger (STS-6)	April 4, 1983	Weitz Bobko Peterson Musgrave	Launched first Tracking and Data Relay Satellite (TDRS-1)
Soyuz T-8	April 20, 1983	Titov Strekalov Serebrov	Unsuccessful docking with Salyut 7
Challenger (STS-7)	June 18, 1983	Crippen Hauck Fabian Ride Thagard	First flight of U.S. woman astronaut, Ride; deployed two communications satellites
Soyuz T-9	June 28, 1983	Lyakhov Aleksandrov	Mission to Salyut 7; crew spent 149 days in space
Challenger (STS-8)	August 30, 1983	Truly Brandenstein Bluford Gardner Thornton	First flight of African-American astronaut, Bluford

Piloted Spaceflight

Program	Launch Date	Crew	Notable Events
Columbia (STS-9)	November 28, 1983	Young Shaw Garriott Parker Lichtenberg Merbold	First spacelab mission; first non-U.S. crew member on a U.S. mission (Merbold)
Challenger (STS-41B)	February 3, 1984	Brand Gibson McCandless McNair Stewart	First in-space operation of Manned Maneuvering Unit (MMU)
Soyuz T-10	February 8, 1984	Kizim Solovev Atkov	Crew spent record-setting 237 days in space
Soyuz T-11	April 3, 1984	Malyshev Strekalov Sharma	First flight of cosmonaut from India
Challenger (STS-41C)	April 6, 1984	Crippen Scobee Hart van Hoften Nelson	Crew launched Long Duration Exposure Facility and repaired Solar Maximum Mission Satellite
Soyuz T-12	July 17, 1984	Dzhanibekov Savistskaya Volk	Savitskaya became first woman to perform EVA
Discovery (STS-41D)	August 30, 1984	Hartsfield Coats Mullane Hawley Resnick Walker	Space Shuttle *Discovery*'s first voyage
Challenger (STS-41G)	October 5, 1984	Crippen McBride Leestma Ride Sullivan Garneau Scully-Power	First flight of two U.S. women, Sullivan and Ride; first Canadian astronaut (Garneau)
Discovery (STS-51A)	November 8, 1984	Hauck Walker Allen Fisher Gardner	Crew retrieved and returned Westar 6 and Palapa B2 to Earth

Piloted Spaceflight

Program	Launch Date	Crew	Notable Events
Discovery (STS-51C)	January 24, 1985	Mattingly Shriver Buchli Onizuka Payton	Mission for U.S. Department of Defense; first Asian-American astronaut (Onizuka)
Discovery (STS-51D)	April 12, 1985	Bobko Williams Seddon Griggs Hoffman Walker Garn	Carried two communications satellites; first flight of U.S. elected official, Senator Garn
Challenger (STS-51B)	April 29, 1985	Overmyer Gregory Lind Thagard Thornton van den Berg Wang	Spacelab-3 mission
Soyuz T-13	June 5, 1985	Dzhanibekov Savinykh	Repair mission to Salyut 7
Discovery (STS-51G)	June 17, 1985	Brandenstein Creighton Fabian Lucid Nagel Al-Saud Baudry	Carried three communications satellites
Challenger (STS-51F)	July 29, 1985	Fullerton Bridges England Henize Musgrave Acton Bartoe	Spacelab-2 mission
Discovery (STS-51I)	August 27, 1985	Engle Covey Fisher Lounge van Hoften	Crew launched three satellites and repaired the disabled Syncom IV-3
Soyuz T-14	September 17, 1985	Vasyutin Grechko Volkov	Boarded Salyut 7; mission cut short due to illness

Piloted Spaceflight

Program	Launch Date	Crew	Notable Events
Atlantis (STS-51J)	October 3, 1985	Bobko Grabe Hilmers Stewart Pailes	First flight of Space Shuttle *Atlantis*; mission for U.S. Department of Defense
Challenger (STS-61A)	October 30, 1985	Hartsfield Nagel Bluford Buchli Dunbar Furrer Messerschmid Ockels	Carried the German spacelab D-1
Atlantis (STS-61B)	November 26, 1985	Shaw O'Connor Cleave Ross Spring Neri-Vela Walker	Three communications satellites launched; first flight of Mexican astronaut
Columbia (STS-61C)	January 12, 1986	Gibson Bolden Chang-Diaz Hawley G. Nelson Cenker B. Nelson	One communications satellite launched; first Hispanic-American astronaut (Chang-Diaz); first flight of member of House of Representatives (B. Nelson)
Challenger (STS-51L)	January 28, 1986	Scobee Smith McNair Onizuka Resnik Jarvis McAuliffe	Shuttle exploded 73 seconds into flight; all seven crew members died; crew included first teacher in space
Soyuz T-15	March 13, 1986	Kizim Solovyov	First occupation of Mir space station; crew transfer to Salyut 7
Soyuz TM-2	February 5, 1987	Romanenko Laveykin	Romanenko set space endurance record of 326 days aboard Mir
Soyuz TM-3	July 22, 1987	Viktorenko Aleksandrov Faris	Mission to Mir; first flight of Syrian cosmonaut
Soyuz TM-4	December 21, 1987	Titov Manarov Levchenko	Mission to Mir

Piloted Spaceflight

Program	Launch Date	Crew	Notable Events
Soyuz TM-5	June 7, 1988	Savinykh Solovyev Aleksandrov	Mission to Mir; crew returned in Soyuz TM-3
Soyuz TM-6	August 29, 1988	Lyakhov Polyakov Mohmand	Mission to Mir; first flight of Afghani cosmonaut
Discovery (STS-26)	September 29, 1988	Hauck Covey Hilmers Lounge Nelson	First U.S. Shuttle flight since *Challenger* accident; launch of TDRS-3
Energiia-Buran	November 15, 1988	unpiloted	First and only flight (to date) of Soviet space shuttle
Soyuz TM-7	**November 26, 1988**	**Volkov Krikalev Chrétien**	**Titov and Manarov remained on Mir for record-setting 366 days**
Atlantis (STS-27)	December 2, 1988	Gibson Gardner Mullane Ross Shepherd	Mission for U.S Department of Defense
Discovery (STS-29)	March 13, 1989	Coats Blaha Bagian Buchli Springer	Launch of TDRS-4
Atlantis (STS-30)	May 4, 1989	Walker Grabe Cleave Lee Thagard	Launched Venus Orbiter *Magellan*
Columbia (STS-28)	August 8, 1989	Shaw Richards Adamson Brown Leestma	Mission for U.S. Department of Defense
Soyuz TM-8	September 5, 1989	Viktorenko Serebrov	Five-month mission to Mir

Piloted Spaceflight

Program	Launch Date	Crew	Notable Events
Atlantis (STS-34)	October 18, 1989	Williams McCulley Baker Chang-Diaz Lucid	Launched *Galileo* probe to Jupiter
Discovery (STS-33)	November 22, 1989	Gregory Blaha Carter Musgrave Thornton	Mission for U.S. Department of Defense
Columbia (STS-32)	January 9, 1990	Brandenstein Wetherbee Dunbar Ivins Low	Crew retrieved Long-Duration Exposure Facility; Syncom IV-5 launched
Soyuz TM-9	February 11, 1990	Solovyov Balandin	Mission aboard Mir for nearly 179 days
Atlantis (STS-36)	February 28, 1990	Creighton Casper Hilmers Mullane Thuot	Mission for U.S. Department of Defense
Discovery (STS-31)	**April 24, 1990**	**Shriver Bolden Hawley McCandless Sullivan**	**Hubble Space Telescope (HST) launched**
Soyuz TM-10	August 1, 1990	Manakov Strekalov	130-day mission aboard Mir
Discovery (STS-41)	October 6, 1990	Richards Cabana Akers Melnick Shepherd	Launch of Ulysses spacecraft
Atlantis (STS-38)	November 15, 1990	Covey Culbertson Gemar Meade Springer	Mission for U.S. Department of Defense

Piloted Spaceflight

Program	Launch Date	Crew	Notable Events
Columbia (STS-35)	December 2, 1990	Brand Gardner Hoffman Lounge Parker Durrance Parise	ASTRO-1 spacelab mission
Soyuz TM-11	December 2, 1990	Afanasyev Manarov	175-day mission aboard Mir
Atlantis (STS-37)	April 5, 1991	Nagel Cameron Apt Godwin Ross	Launch of Compton Gamma-Ray Observatory (GRO)
Discovery (STS-39)	April 28, 1991	Coats Hammond Bluford Harbaugh Hieb McMonagle Veach	Mission for U.S. Department of Defense
Soyuz TM-12	May 18, 1991	Artsebarsky Krikalev Sharman	Mission to Mir; first flight of cosmonaut from United Kingdom
Columbia (STS-40)	June 5, 1991	O'Connor Gutierrez Bagian Jernigan Seddon Gaffney Hughes-Fulford	Spacelab Life Sciences (SLS-1) mission
Atlantis (STS-43)	August 2, 1991	Blaha Baker Adamson Low Lucid	Launched TDRS-5
Discovery (STS-48)	September 12, 1991	Creighton Reightler Brown Buchli Gemar	Upper Atmosphere Research Satellite launched
Soyuz TM-13	October 2, 1991	Volkov Aubakirov Viehboeck	Mission to Mir

Piloted Spaceflight

Program	Launch Date	Crew	Notable Events
Atlantis (STS-44)	November 24, 1991	Gregory Henricks Musgrave Runco Voss Hennen	Defense Support Program satellite launched
Discovery (STS-42)	January 22, 1992	Grabe Oswald Thagard Hilmers Readdy Bondar Merbold	International Microgravity Laboratory-1 (IML-1) Spacelab mission
Soyuz TM-14	March 17, 1992	Viktorenko Kaleri Flade	First piloted mission for the former Soviet Union
Atlantis (STS-45)	March 24, 1992	Bolden Duffy Sullivan Foale Leestma Frimout Lichtenberg	Atmospheric Laboratory for Applications and Science (ATLAS) Spacelab mission
Endeavour (STS-49)	May 7, 1992	Brandenstein Chilton Akers Hieb Melnick Thornton Thuot	First flight of Space Shuttle *Endeavour*; retrieved Intelsat VI
Columbia (STS-50)	June 25, 1992	Richards Bowersox Dunbar Baker Meade DeLucas Trinh	U.S. Microgravity Laboratory-1 Spacelab mission
Soyuz TM-15	July 27, 1992	Solovyov Avdeyev Tognini	Mission to Mir

Piloted Spaceflight

Program	Launch Date	Crew	Notable Events
Atlantis (STS-46)	July 31, 1992	Shriver Allen Hoffman Chang-Diaz Ivins Nicollier Malerba	Launched EURECA for European Space Agency; tested Tethered Satellite System
Endeavour (STS-47)	September 12, 1992	Gibson Brown Lee Apt Davis Jemison Mohri	Jemison first African-American woman in space; Japanese Spacelab-J mission
Columbia (STS-52)	October 22, 1992	Wetherbee Baker Jernigan Shepherd Veach MacLean	Launch of Lageos II; first flight of U.S. Microgravity Payload (USMP)
Discovery (STS-53)	December 2, 1992	Walker Cabana Bluford Clifford Voss	Launch of U.S. Department of Defense satellite
Endeavour (STS-54)	January 13, 1993	Casper McMonagle Harbaugh Helms Runco	Launch of sixth TDRS satellite; failed to reach correct orbit
Soyuz TM-16	January 24, 1993	Manakov Poleshchuk	Mission to Mir; test U.S. Shuttle docking unit
Discovery (STS-56)	April 8, 1993	Cameron Oswald Cockrell Foale Ochoa	Second ATLAS Spacelab mission; launch of Spartan probe
Columbia (STS-55)	April 26, 1993	Nagel Henricks Ross Harris Precourt Schlegel Walter	Spacelab D2 mission for microgravity research

Piloted Spaceflight

Program	Launch Date	Crew	Notable Events
Endeavour (STS-57)	June 21, 1993	Grabe Duffy Low Sherlock Voss Wisoff	Launched and retrieved European satellite EURECA; first flight of Spacelab module
Soyuz TM-17	July 1, 1993	Tsibliyev Serebrov Haignere	Mission to Mir
Discovery (STS-51)	September 12, 1993	Culbertson Readdy Bursch Newman Walz	Launched and retrieved European satellite ORFEUS-SPAS
Columbia (STS-58)	October 18, 1993	Blaha Searfoss Seddon Lucid McArthur Wolf Fettman	First two-week Shuttle mission; second Spacelab Life Science (SLS-2) mission
Endeavour (STS-61)	December 2, 1993	Covey Bowersox Musgrave Akers Hoffman Nicollier Thornton	Crew serviced and repaired Hubble Space Telescope

Special thanks to Althea Washington at NASA Headquarters and George Michael Gentry at Johnson Space Center.

NASA—National Aeronautics and Space Administration
SI—Smithsonian Institution
NASM—National Air and Space Museum
KSC—Kennedy Space Center
ESA—European Space Agency
GSFC—Goddard Space Flight Center

Photo sources and negative numbers are indicated.

1 NASA S80-36190; **2–3** NASA 73-HC-789; **4–5** NASA 51L-(S)-157; **6–7** NASA AS11-40-5903; **8** NASA S65-63220; **9** SI, NASM 80-3070; **11** Martin Marietta; **15** NASA 81-HC-4; **16** NASA 68-HC-132 (l), SI, NASM A-4086-E (r); **17** SI, NASM 84-14155; **18** NASA 92-HC-356; **19** NASA STS-047-37-003; **20** NASA 91-HC-451; **21** Sovfoto D-9181; **23** NASA KSC-69PC-422 (l), NASA KSC-69PC-417 (r); **24** NASA S34-(S)-063 (l), NASA 83-HC-451 (r); **25** NASA S29-(S)-063; **26** NASA 71-HC-1433; **27** NASA S29-78-003 (t), Martin Marietta (b); **28** Earth Data Analysis Center (t), NASA S42-22-006 (b); **29** NASA 68-HC-641; **30** NASA 69-HC-762; **31** NASA 67-HC-745; **32–33** NASA 65-HC-1260; **34** NASA 71-HC-1002; **35** NASA 71-HC-1012 (l), NASA 85-HC-150 (r); **37** Princeton University Library, courtesy of SI, NASM 78-10190; **38** Science Museum, London, courtesy of SI, NASM 84-1793; **39** SI, NASM 94-6828 (l), SI, NASM 94-8268 (r); **40** SI, NASM 75-11483 (t), SI, NASM 76-17287 (b); **41** NASA 74-H-1220, courtesy of SI, NASM; **42** SI, NASM A-42103 (l), NASA 74-H-1202, courtesy of SI, NASM (r); **43** SI, NASM 90-16503; **44** SI, NASM 77-11216; **45** SI, NASM 76-7558 (t), SI, NASM 76-4433 (b); **46** K. E. Tsiolkovsky State Museum of Astronautics, courtesy of SI, NASM 73-7133 (l), TASS from Sovfoto D-30483 (r); **47** SI, NASM 76-17275; **48** Marshall Space Flight Center; **49** SI, NASM 91-9174 (t), SI, NASM 86-13268 (b); **50** SI, NASM 76-7559; **51** NASA 69-HC-761 (l), U.S. Space & Rocket Center, courtesy of SI, NASM 91-19620 (r); **53** NASA 82-HC-219; **54** General Dynamics (l), NASA 62-MA8-56 (r); **55** NASA Telstar 13; **56** Martin Marietta; **57** NASA 69-HC-622 (l), NASA 73-HC-436 (r); **58–59** KSC-392C-2230.35 (l), KSC-392C-2230.05 (c), KSC-392C-2232.03 (r); **60** NASA 81-HC-46 (l), NASA 72-HC-858 (b); **61** NASA 82-HC-360; **62** NASA 79-HC-540; **63** NASA S26-31-036; **64** Sovfoto D-147747 (l), Sovfoto D-54475 (r); **65** Sovfoto D-30742; **66** Sovfoto D-98983; **67** ESA; **68–69** Han Xinhua/Sygma; **70** National Space Development Agency of Japan (l, r); **72–73** McDonnell Douglas (l, c, br), NASA 76-HC-718 (tr); **75** NASA S83-35801; **76** Sovfoto D-580; **77** SI, NASM 75-10226 (l), Sovfoto D-600 (r); **78** Jet Propulsion Laboratory 7-3; **79** NASA 59-EX 28-VII (l), NASA 68-HC-383 (r); **80** NASA VAN-II; **81** Sovfoto D-N-71-1012; **82** SI, NASM 74-12209; **83** SI, NASM 73-379; **84** SI, NASM 94-7710 (l), SI, NASM 94-7709 (r); **85** NASA 75-HC-237; **86** NASA 75-HC-606 (l), NASA AST 1-056 (r); **87** NASA 75-HC-230; **89** NASA MA9-83; **90** NASA MR-3-8; **91** NASA 62-MA6-108; **92–93** NASA MA6-89 (l), NASA S62-6029 (c), NASA 62-MA6-178 (r); **94** NASA 65-HC-958; **95** NASA 65-HC-294; **96** NASA 65-HC-363; **97** NASA 65-HC-303; **98** NASA 65-HC-1261; **99** NASA 66-HC-99 (l), NASA 65-HC-2158 (c), NASA 66-HC-354 (r); **100** NASA 66-HC-911; **101** NASA 66-HC-1886 (t), NASA 66-HC-1878 (b); **102** Sovfoto D-21019; **104** NASA SL3-114-1683; **105** NASA 73-HC-743 (l), NASA 73-HC-228 (r); **106** NASA 73-HC-470 (t), NASA 73-HC-242 (b); **107** NASA 73-HC-520; **108** NASA S84-27024; **109** NASA 73-HC-742 (t), NASA 76-HC-8 (b); **110–111** NASA AST-2-093 (l), NASA AST-5-298 (c), NASA AST-32-2686 (r); **113** Sovfoto D-29635; **114** NPO Energiia; **115** Sovfoto D-47192;

116 Sovfoto D-49802; **118–119** NASA 89-HC-578 (l), NASA 83-HC-466 (r); **120** NASA 81-HC-763; **121** NASA 84-HC-57 (l), NASA 82-HC-730 (r); **122–123** NASA 80-HC-40; **124** NASA 61B-41-019 (t), NASA STS050-21-035 (b); **125** NASA S19-39-034 (l), NASA S13-37-1718 (r); **126** NASA 90-HC-299; **127** NASA STS047-39-034 (l), NASA 83-HC-240 (r); **128** NASA 84-HC-75 (t), NASA 61B-102-0022 (b); **129** NASA 84-HC-74; **130** NASA S82-33421; **131** Sovfoto D-98484; **133** NASA/GSFC AS15-85-11435; **134** SI, NASM 81-12136; **135** Sovfoto D-N-1002 (t), NASA 66-HC-899 (b); **136** John F. Kennedy Library (l), NASA 71-HC-611 (r); **137** NASA S70-30534 (t), NASA 65-HC-959 (b); **138** NASA 69-HC-620; **139** NASA 72-HC-978; **141** NASA S69-19197 (l), NASA AS15-88-11974 (r); **142** NASA 69-HC-617; **144** NASA AS14-66-9306; **145** NASA AS9-20-3064; **146** NASA/GSFC AS11-44-6551 (t), NASA S69-39962 (b); **147** NASA S69-42583 (t), NASA AS11-40-5878 (b); **148** NASA AS11-40-5903 (l), NASA S69-40308 (r); **149** NASA 69-HC-916; **150** NASA 70-HC-501 (l), NASA 70-HC-467 (r); **151** NASA/GSFC AS14-68-9405; **152–153** NASA 69-HC-153; **154** NASA AS15-86-11603; **157** NASA/GSFC AS17-140-21497; **159** NASA 74-HC-655; **160** NASA 80-HC-641; **161** NASA S74-23266; **162** SI, NASM A-391 D; **163** Sovfoto D-N-86-58 (l), Sovfoto D-30737 (r); **164** NASA S88-51854; **165** Martin Marietta; **166** NASA 92-HC-569; **167** SI, NASM 82-2149; **168** Sovfoto 39253; **169** NASA 75-H-523 (l), NASA 71-HC-664 (r); **170** NASA 76-HC-660; **171** NASA 76-HC-855 (t), SI, NASM 80-3070 (b); **172** SI, NASM 80-4979; **173** NASA S93-30705; **175** NASA 74-HC-152 (t), NASA 72-HC-42 (b); **176** NASA 72-HC-133; **177** NASA 79-HC-66 (t), NASA P-24653 (b); **179** NASA 77-HC-333; **180** NASA S89-42091 (l), NASA S89-48714 (r); **181** Martin Marietta; **182** ESA (l, r); **183** NASA 86-HC-144; **185** NASA S19-41-063; **186** NASA 84-HC-418 (t), NASA S64-31395 (b); **187** Martin Marietta; **188** NASA S49-52-025; **191** Rockwell International (t), Martin Marietta (b); **192** NASA KSC-94PC-583; **193** NASA 83-HC-74 (l), General Dynamics (r); **194** NASA 93-HC-305 (l), NASA 86-HC-291 (r); **195** NASA 89-HC-275; **196** NASA 94-HC-2; **197** NASA 94-HC-21; **198** NASA 79-HC-114 (t), NASA 67-HC-772 (b); **199** NASA 77-HC-32; **201** NASA 92-HC-20; **202** NASA 84-HC-267; **203** NASA 92-HC-251; **204** Lawrence Livermore National Laboratory; **205** NASA 91-HC-372 (l), Lawrence Livermore National Laboratory (r); **207** NASA S94-31182; **208** NASA 91-HC-418. Glossary: **Aldrin** NASA 69-H-969, courtesy of SI, NASM; **Apollo 11 and 13** NASA 71-H-534; **Apollo 15** NASA 71-H-1232; **Apollo 17** NASA 72-H-1254; **Armstrong** NASA S66-24453, courtesy of SI, NASM; **Borman** NASA 65-H-1971, courtesy of SI, NASM; **Centaur** NASA 61-C-5; **Collins** NASA 66-H-967, courtesy of SI, NASM; **Haise** NASA S69-62238, courtesy of SI, NASM; **Kohoutek** NASA 74-H-76; **Marisat B** NASA 76-H-453; **Delta** NASA 82-H-728; **crawler** NASA 65-H-678; *Endeavour* NASA 92-H-447; **ERBS** NASA 84-H-628; **Gagarin** SI, NASM A-393-E; **Goddard** NASA 74-H-1246, courtesy of SI, NASM; **Grissom** NASA S65-19552, courtesy of SI, NASM; **Korolëv** SI, NASM 76-17276; **Komarov** SI, NASM 73-7159; **Pegasus rocket** NASA 89-H-425; **Lovell** NASA 70-H-618, courtesy of SI, NASM; **GOES-G** NASA 87-H-204; **McDivitt** NASA 63-Astronaut-43, courtesy of SI, NASM; **SEASAT** NASA 78-H-233; **Oberth** SI, NASM A-979; **Ranger B** NASA 63-Ranger B-1; **rendezvous simulator** NASA 63-Gemini-2; **Titan-Centaur** NASA 73-H-1274; **Sputnik 2** SI, NASM 88-10519; **Skylab 2** NASA 73-H-580; **Schirra** NASA S65-56197, courtesy of SI, NASM; **Schweickart** NASA S64-35343, courtesy of SI, NASM; **Soyuz** TASS/Sovfoto D-N-1031; **Stafford** NASA S66-42576, courtesy of SI, NASM; **Shepard** NASA S64-35343; **Tereshkova** SI, NASM 83-2837; **Titov** SI, NASM 94-8425; **Tsiolkovsky** SI, NASM A-4110-A; **von Braun** SI, NASM 76-13637; **von Opel** SI, NASM 94-8419; **weightlessness** NASA 91-H-385; **White** NASA 64-H-2670, courtesy of SI, NASM; **Young** NASA 69-H-648, courtesy of SI, NASM.

Spaceflight on Display

**The National Air and Space Museum,
Smithsonian Institution**
6th St. & Independence Ave., SW
Washington, DC 20560
(202) 357-1400 or (202) 357-2700
Housing the definitive collection of artifacts from
the U.S. space program, the national museum
preserves and exhibits the most significant vehicles
in the history of spaceflight. Collection includes
most flown U.S. piloted spacecraft. Displays include
the Gemini IV capsule, the Apollo 11 command
module and a lunar lander, Moon rocks, Skylab,
Apollo-Soyuz, and many rockets ranging from
Goddard rockets and a V2 to a Minuteman III and
a Soviet SS-20 ICBM.

The U.S. Space & Rocket Center
One Tranquility Base
Huntsville, AL 35807
(205) 837-3400
Showcased are Mercury, Gemini, and Apollo
spacecraft, a Lunar Module, a Saturn V rocket, and
a full-scale mock-up of the Space Shuttle.

**Space Center Houston
NASA Lyndon B. Johnson Space Center**
Houston, TX 77058
(713) 244-2105
Visitors can take a "behind the secnes" tram tour of
astronaut training and mission control facilities,
take control of the Space Shuttle in a computer
program, and enjoy IMAX films of spaceflight. On
display are flown Mercury, Gemini, and Apollo
spacecraft; a Saturn V, a lunar rover, a Moon rock,
the Skylab trainer, and many spacesuits.

Spaceport USA
NASA John F. Kennedy Space Center
Cape Canaveral, FL 32899
(407) 452-2121
Exhibits at this major launch site for military and
civilian missions include rockets, memorabilia,
artifacts, equipment, and hands-on displays. A full-
scale Space Shuttle model is showcased.

Other NASA visitor centers include:
NASA Goddard Flight Center
Goddard Space Flight Center, Code 130
Greenbelt, Maryland 20771
(301) 286-8981

NASA Langley Research Center
Mail Stop 480
Hampton, VA 23665
(804) 864-6000

NASA Lewis Research Center
21000 Brookpark Rd.
Cleveland, OH 44135
(216) 433-2001

NASA Wallops Flight Facility
Wallops Island, VA 23337
(804) 824-2298

More About Spaceflight

Bond, Peter. *Heroes in Space: From Gagarin to
Challenger*. New York: Basil Blackwell, Inc. 1989.

Braun, Wernher von, Frederick I. Ordway III, and
David Dooling. *Space Travel: A History*. New
York: Harper & Row Publishers, Inc. 1985.

Chaikin, Andrew. *A Man on the Moon: The
Voyages of the Apollo Astronauts*. New York:
Viking, 1994.

Joels, Kerry Mark and Gregory P. Kennedy. *The
Space Shuttle Operator's Manual*. New York:
Ballantine Books, 1982.

McDougall, Walter A. *The Heavens and the Earth:
A Political History of the Space Age*. New York:
Basic Books, 1985.

Murray, Bruce. *Journey into Space: The First Three
Decades of Space Exploration*. New York: W.W.
Norton & Company, Inc., 1989.